On Human Geography

for Rita

R. J. Johnston

On Human Geography

Basil Blackwell

© R. J. Johnston 1986

First published 1986
First published in paperback in 1988

Basil Blackwell Ltd
108 Cowley Road, Oxford OX4 1JF, UK

Basil Blackwell Inc.
432 Park Avenue South, Suite 1503
New York, NY10016, USA

British Library Cataloguing in Publication Data

Johnston, R. J.
    On human geography.
    1. Anthropo-geography
    I. Title
    304.2   GF41
    ISBN 0-631-14023-9
    ISBN 0-631-14024-7 (Pbk)

Library of Congress Cataloging in Publication Data

Johnston, R. J. (Ronald John)
    On human geography.
    Bibliography: p.
    1. Anthropo-geography. 2. Geography. I. Title.
    GF41.J63 1986   304.2   85–30800
    ISBN 0-631-14023-9
    ISBN 0-631-14024-7 (pbk.)

Typeset by Oxford Publishing Services, Oxford
Printed in Great Britain by Billing & Sons, Worcester

Contents

Preface   vi

1 Introduction   1

2 A Geographical Context   6

3 A Model of Society   21

4 Story-Telling: or Doing Geography   50

5 Positivism, Science and Quantification   83

6 Applied and Applicable   100

7 Geographical Fixations   122

8 A Human Geographical Education   144

9 Moving Forward   168

Bibliography   182

Index   194

Preface

This book will be published close to the twenty-second anniversary of my taking up my first academic appointment. Already, a reviewer (working in the same department that I joined in 1964) has suggested that I am going through a mid-career crisis. That crisis has probably always been latent within my career, but it is only recently that I have perceived, relatively clearly, the nature of the shift that is necessary to its resolution. The present book is a statement of that perception.

I am grateful to John Davey for encouraging me to write the book and for his solicitous oversight of its production. Once again, I am grateful to Joan Dunn for her work on the manuscript. Above all, my thanks are due to Rita, constant companion and support during the years of crisis.

1

Introduction

I can only make direct statements, only 'tell stories', whether or not
the stories are 'true' is not the problem. The only question is whether
what I tell is *my* fable, *my* truth.

Carl Jung, Prologue to Memories, Dreams, Reflections, 1953

In recent years I have produced two books – *Geography and Geographers:
Anglo-American Human Geography since 1945* (Johnston, 1983a) and
*Philosophy and Human Geography: An Introduction to Contemporary
Approaches* (Johnston, 1983b) – which have explored recent philosophical
and methodological debates within geography. They were explicitly
produced as texts, syntheses of a large literature that emerged from my own
teaching of these subjects, and were designed to provide an overview for
others. They were neither neutral nor objective, for that is impossible, but in
them my own opinions were never explicit; the books were my interpretations of the literature, however, both in how I read the individual pieces and
how I put the pieces together. This led to comments by some reviewers that
the books were rather detached and apart from the debates, and that my
personal involvement was hard to capture (though see Mercer, 1985b).

The present book is in part a response to those comments, but also to
other stimuli, including my reflections on those two books, on many other
aspects of my career, and on my life. In producing *Geography and
Geographers* (the first edition was published in 1979) I was initially
attracted to the paradigm model of disciplinary development and thought
that the period I was describing could be neatly encapsulated into a series of
blocks, separated by some interesting revolutions. By the time I had
assembled the material I was unconvinced, but thought that a modified
version of the model would work. When I produced the revised version of
the book four years later I was convinced that even the modified model was
untenable, and said so. (Nevertheless, the book is still imbued with it, and if
there is a third edition, much more rewriting will be needed.) Realization of
this came in two ways. First, as I read and reread more recent human

geography I appreciated the futility of classifying, in any grand sense, the great variety of work being done contemporaneously; I was shifting more and more towards an anarchic model, of disciplinary communities responding in their own contexts to the desire to practise as geographers. Secondly, I had been asked to provide an academic autobiography for *Recollections of a Revolution* (Billinge, Gregory and Martin, 1984) and in reviewing my own career I became aware of the importance of context – not only physical but also bibliographic. I was, I realized, 'A foundling floundering in World Three'; if others experienced similar paths (and other essays in that book clearly indicate that they did) then it is surprising that I was able to structure the material for *Geography and Geographers* at all. The chemistry of a discipline is not so easy to discern.

Preparing *Philosophy and Human Geography* strengthened these conclusions. Again, I thought that there was a simple structure, a set of compartments into which work could be categorized. In general terms, I believe I was right, and that for heuristic purposes the compartments that I chose were sensible. But the actual task of categorization was much harder than I anticipated (to stick to the paradigm model, there were too many anomalies), and my attempt to provide a structure is thus in part an oversimplification. Most important of all, however, was the writing of the final chapter to that book on 'Conflict and accommodation'. I concluded by stating that:

> This book has ended with a dilemma, deliberately so, because human geographers face a dilemma at the present time . . . They must either choose between a variety of approaches to human geography or seek to fashion a middle course incorporating elements of two or more of those approaches. Further, they must choose in the knowledge that the concepts of value-freedom and objectivity in social science are under increasing challenge, as it is demonstrated that theory and practice are inseparably linked. Their choice has political as well as academic connotations, which they cannot avoid. (p. 134)

That dilemma was my dilemma; that choice was my choice. I had been struggling with it for some time, desperately trying to fashion a satisfactory personal approach to human geography (Johnston, 1986a). My biggest problem was finding some accommodation for the empirical work that I was enjoying doing – the quantitative description of voting patterns, the analysis of the geography of public policy in the USA and, most enjoyable of all, the unravelling of the legal complexities of suburban exclusiveness in the USA – and the general models (I called them structuralist then, and still do now) of society which I found attractive. The root of the problem, I now realize, was that I falsely equated empirical research with positivism. Once I was able to break that false equation, I could see a way forward. This book is about that way.

The break was brought about by a continued floundering in 'World Three' and by reflection on the whole of my life, rather than, as before, separating out Ron Johnston as academic geographer from Ron Johnston as husband, father, Sheffield resident, change-ringer, supporter of Swindon Town and Geoffrey Boycott etc. etc. In this way I realized the the problem was not empirical research but empiricism; the former can open my eyes and mind, but the latter closes them. From then on, the task was still difficult, but at least I knew where I was going.

The present book describes where I have got to. It is not presented as a text or as a manifesto. Later on (p. 165) I present a classification of academic books. Within that classification, this book is clearly an essay, a piece written to stimulate interest. It is certainly not a textbook; some might call it an indulgence.

The book begins by providing a brief contextual introduction. The dissatisfactions that I had with my own practice of geography arose very largely out of the problems I had accommodating what I was doing with my general interpretations of the world around me: the world I lived in did not seem to be the same place as the world I studied. I needed to bring the two together, and it was my reflections on 'the world outside', together with my continued reading as an academic, that pushed me towards the position I now occupy. Chapter 2 provides a short survey of those reflections, to indicate their importance in the direction that I have taken and which is the major contribution of this book. It is in no sense an autobiography, though (as John Eyles, 1985, has recently shown) autobiography can be a good source of geographical awareness.

The most important conclusion reached from the considerations reported in chapter 2 is the need for a viable theory of society in which to set my work. Chapter 3 is an outline of that theory. Its basic features are: that it is realist, based on an attempted understanding of the materialist base of society; that it is neither determinist nor voluntarist – people are neither 'puppets' nor 'free agents' but individuals taking decisions in contexts over which they have little personal control; and that the subject matter of geography (space, place and environment) are integral parts of that theory. That chapter 3 is only an outline, not a fully articulated statement, has to be stressed: it is a framework only.

Given the framework, the next pressing problem is how to use it. As mentioned above, a major difficulty that I experienced was in the relationship between my developing realist theory and the conduct of empirical research. Here, further reflection on the material in *Philosophy and Human Geography* came to the rescue, and I was able to see that many of the debating points about philosophical issues were irrelevant to the conduct of human geography as social science. The goal is rigorous (scientific) evaluation of evidence to provide a coherent understanding of the empirical world and of how people act in it. They act by conjointly interpreting the driving forces in society (the mechanisms in the realist philosophy) and the context in which they reside, and my goal is to understand how they do that, to produce a coherent portrayal of their actions, coherent that is with both what they do and the theoretical context in which I am setting the portrayal. A viable understanding – a good story – is one that faithfully portrays both. The theory, if it is to be valid, must provide a framework within which I can understand the empirical material.

Chapters 4 and 5 are presentations of the art of geographical story-telling that I have developed. In the first, I deal with the general issues of how to 'do geography' within a realist framework. Chapter 5 is a by-product of that. Much of the debating in human geography in recent years has ranged over the philosophy of positivism. A lot of it is ill-informed, since it equates positivism with both scientific rigour and quantification. Chapter 5 seeks to dispel some of the misunderstandings, arguing that neither scientific method as the rigorous evaluation of hypotheses nor quantification is outlawed from the sort of human geography that I want to practise.

Harrison and Livingstone (1980) have argued that social conscience and philosophical awareness cannot be separated, so that the presuppositions with which we begin research are coloured by our fundamental beliefs about the origin of reality. These are our cosmologies, our personal paradigms, that condition our philosophies, our beliefs about the sources of knowledge and about knowledge itself. Only when our cosmologies and philosophies are determined can we shift to a consideration of methodologies, means of obtaining knowledge. Thus chapter 3 presents the cosmology, chapter 4 the philosophy, and chapter 5 the methodology of my approach to human geography.

Chapters 2 to 5 are the core of the book, therefore. The remaining chapters take up important issues linked to that core, but not necessarily a linear progression from it; they are about aspects of the practice of human geography. Chapter 6 is concerned with the crucial issue of the uses of science. As I make clear there, calls for human geography to be 'relevant', to be strongly applied in its orientation, are commonplace today. All science is, of course, applicable – for good or evil – but the version of applied science being promoted is a limited one, linked to a particular philosophy, that of empiricism/positivism. Retaining the categories of science first presented in *Philosophy and Human Geography*, I argue that there are three types of applied science and thus three types of applied geography. All are valid. All, I believe, are needed; there are difficulties in choosing one over another because they raise a crucial human issue: can any individual's life chances today be sacrificed for those of somebody else tomorrow?

Chapter 7 is concerned with the internal structure of the discipline of human geography and its links with other disciplines. In seeking a rationale for themselves and their discipline, geographers have advanced a range of manifestos, and several of these have become what I see as major geographical fixations, or fetishisms. I deal with four of these, evaluating the cases for: the integration of physical and human geography; human geography as spatial science; regional geography; and systematic geography.

In chapter 8 I turn to questions relating to the organization of the practice of human geography, with especial reference to education. This is probably the most pessimistic of the chapters. Having developed my cosmology-philosophy-methodology, and linked this to a clear view of how science can be applied, I evaluate educational practice (British only) at both school and university/polytechnic/college level. I find that a limited conception of human geography is being presented, and – more depressingly – that because of the forces within society this is unlikely to change. Human geography should be liberating, illuminating the world and freeing people to act constructively in it, but this is not happening.

Depression must not lead to defeatism, and the final chapter is thus entitled Moving Forward. This is because the book is, for me, only a start, a preface to the next phase in my academic career. How to carry my ideas forward, how we should practise geography, are briefly considered there.

This then is a book written by me and for me, as a formalization of the point I have reached and a signpost for the direction I want to take. It is presented to others in the hope that it will be of benefit for them too – in a positive and not a negative sense, I hope – in charting their route forward. Clearly I am privileged to be able to do this, to seek to influence how people structure the human geography of the future – both the world that is studied and the way that it is studied. Having struggled to create some coherence out of the many issues that have concerned human geographers in recent years, this statement is presented as an aid, to motivate others seeking to do likewise.

2

A Geographical Context

There is a restlessness in our air that makes people dissatisfied with what they have got, and makes them want not *more* things but *other* things. Such people are not greedy; they are simply trying to fill an unrecognized nonmaterial want with a safely intelligible material one ... What we need is an advertising campaign to sell, instead of novelties, oldities – by which I mean not second-hand earring caddies but self-understanding ...

Bernard Levin, The Times, 20 August 1985

For most human geographers, the subject matter of their academic interests cannot be divorced from the content of their daily lives. Their object of study is the world that they live in, and many of them are contributors to the material that they study (I am a British voter, for example, as well as a student of British voting patterns, and I am a part-owner of a house as well as an analyst of residential choices). Some of them may try to separate the two, to allow what they study to have no influence on their 'non-academic' activities and vice versa, but they are unlikely to succeed. Almost certainly, what they read and research will influence them as citizens and how they think and act in their non-working lives will influence what and how they research and teach. This has certainly been the case for me, and has led to much thought about the nature of the academic discipline that I joined. The content of that thinking is outlined here, since it is a major influence on the personal approach to human geography that I develop in the rest of this book.

DEVELOPMENT AND IDEOLOGY

Few citizens of the industrial societies are unaware, at least superficially, of the gross inequalities characteristic of the world today, even if it takes a rock concert extravaganza or similar media event to stimulate any consideration of their implications. As a geographer, my awareness of those inequalities is heightened, yet I would hope that all of my fellow-citizens are aware of how much we are above the global average, not only in terms of material possessions but also, and much more importantly, in terms of general health and life expectancy. I am well-fed, well-housed, and well-clothed, and am likely to be so for the 30 or more years that, according to national averages, I still have to live. But the vast majority of the world's people are poor, hungry, live in appalling conditions, suffer from bad health, and are likely to die very soon.

It is at the global scale that these inequalities are usually presented, and are graphically portrayed in such excellent geography books as Kidron and Segal's *New State of the World Atlas* (1984). Thus the problems inherent in those inequalities are presented, especially in the media, as occurring somewhere else. They are almost universally associated with identified countries, and so presented as national problems – indeed, such a presentation stimulates nationalism. We are, of course, aware of inequalities within our own country, though for many of us such awareness is as partial and transitory as the awareness of global inequalities because we live in very closed, small worlds – we all need to refer to Fothergill and Vincent's *State of the Nation Atlas (1985)*. To geographers, works such as Bill Bunge's (1971; Bordessa and Bunge, 1975) on Detroit and Toronto and Paul Harrison's *Inside the Inner city* (1983) make clear the extent of those intra-national inequalities and show how the use of national averages can conceal more than they reveal, but to many no such eye-opening takes place except perhaps occasionally, and vicariously, by TV.

To some of us, the existence of such inequalities at all spatial scales is a moral issue. We are uneasy about our privileged positions, and as geographers that unease is heightened by the close contact that we may have with sufferers of the allocation systems. We want to correct those inequalities, to make everybody as well-off and as healthy as ourselves. And so we seek to understand how inequalities are created and how they might be eradicated.

Models of Development

Two sets of views of the production of inequalities are made available to me; within each set there are many differences in presentation and argumentation, but the separation of the two is clear. I must choose between them, must decide which portrayal I find the most convincing.

The first set argues that the present organization of production, distribution and exchange in the world is immoral, because it is based on the exploitation of the mass of the world's population and resources in order that a small minority, including me, might be privileged and pampered. That pampering is excessive: I over-eat, have too many clothes and material possessions relative to my real needs, am profligate in my consumption of energy and other resources, and so on. And while I enjoy that situation, and an expected life-span of more than 70 years, most people live in squalor and can expect to live for little more than half that time. I am, then, contributing to structural violence, to the pauperization and early death of many in order that I can indulge my life-style for longer. My feelings of guilt at this realization should ensure that I work for the creation of a new system of production, distribution and exchange which would not require such inequalities and structural violence.

The other set of arguments focuses on a belief that the inequalities are temporary, an unfortunate consequence of the uneven spatial adoption of the methods and disciplines of securing what we know as economic development. (I well remember how this argument entered the human geography literature in the 1970s, when Brian Berry, 1970, promoted Jeffrey Williamson's (1965) findings about regional levelling-up to the status of a universal law.) The lot of everybody will improve substantially, I am told, if we all work hard to make the present system of economic organization effective. There may be some exploitation, with a minority doing better than others, but this is a small price to pay for high average standards – and in any case, I am assured, it is not really exploitation that produces the slight inequalities but just the rewards for above-average amounts of enterprise and hard work. We are all so well-off in the already advanced countries because we have given enterprise free rein and rewarded it, and it is only because we tend now to denigrate enterprise that we are unable to react rapidly to the shifts that it promotes in order to retain our prosperity, hence our current little difficulties. No other societies have ever enjoyed such widespread material benefits or such healthy and long lives. The rewards of free enterprise have been the betterment of us all: we should ensure that this continues, and should encourage others elsewhere to act in exactly the same way.

According to the second set of views, which is much more frequently presented to me (as a British citizen if not as an academic geographer) the path to a better life for all in the world is a continuation and expansion of the free enterprise system. After all, are not black workers in South Africa better off materially than their contemporaries on the rest of that continent because South Africa has the most successful free enterprise economy there? And will they not be even better off if we encourage the development of that economy and stop trying to influence social policies there, to the detriment of economic advancement? South Africa, South Korea, Hong Kong, Taiwan: these are the success stories of the extension of the free enterprise system in recent decades, and point to the needs for the future. Our concern should be with economic development through investment, and this alone will solve many of the other problems. For example, most countries in the Third World currently have high birth rates, and their population expansion is a drain on natural resources, threatening the survival of the global ecosystem. But we used to have high birth rates too. They fell as material standards rose and people were presented with choices in their way of life. The same will happen in the Third World, I am told; the best contraceptive is economic development.

If accompanied by the use of historical analogies, the second set of arguments carries a lot of force, and indicates what my choice should be. The right policy, I would conclude, is to let free enterprise have free rein worldwide so that all countries would pass through Rostow's renowned stages of economic growth, producing a world of equals. But is that the right policy? As I ponder the question, so I am forced to tackle a number of issues.

First, the historical analogy is difficult to sustain. The nineteenth century economic development of Britain was built on the exploitation of cheap labour and resources, at home initially but increasingly overseas as our economic system extended into all four other continents. The development of other countries proceeded likewise, creating a world largely divided into a small number of spheres of influence. But what happened when other countries sought similar expansion? There was nowhere left to colonize, and competition over spheres of influence generated conflict and, eventually, world war. In a finite world, it seemed, the possibilities for development were limited.

If this is the case, then the prospects for economic development in what we know as the Third World must be very limited, for expansion there can only be at another country's loss – and it is very doubtful if the already economically developed (hence militarily and politically powerful) will readily yield spheres of influence. But are spheres of influence necessary? There is a necessary polarization, it seems, between development and underdevelopment. The former, defined as increased GNP per capita, depends on evermore efficient production, which means a relative decline in the costs of labour and materials. Thus if there is no population and/or resources to underdevelop, there can be no development. Within any defined area, the population and resources are finite, and so is the degree of potential underdevelopment. (That potential is also influenced by local resistance to underdevelopment, a point explored later on.) Thus spatial expansion, in the search for cheap labour and materials, is necessary to the development trend. If one area is to develop, another will be underdeveloped – in relative terms, at least, for the latter may have high material standards compared to other, less underdeveloped areas (as the South African case shows). Given the present configuration of economic, political and military power in the world, therefore, it is difficult to see widespread development in the contemporary Third World because of the absence of anywhere to underdevelop.

Secondly, the clear lesson of history is that development under the free enterprise system is not a unilinear sequence. It has proceeded through booms and slumps over the past three or four centuries, as students of Kondratieff and other cycles have shown, and there is no reason to believe that this sequence will change. After every period of economic boom there is a slump, as the free enterprise system is restructured, and even in the developed countries those slumps are marked by widespread (relative) distress. Is a system that makes an occasional generation a scapegoat to its need for periodic restructuring a desirable one – especially if, as some commentators suggest (e.g. Harvey, 1985), the price of recovery may have to be war and widespread slaughter.

Thirdly, and following on from the first two issues, the basis of the free enterprise system is competition. Any competition must have losers as well as winners – the permanent draw is unlikely. And so in a competition on which the contestants' livelihoods depend, the losers may have their way of life, if not life itself, destroyed. The creation of welfare states is advanced as a means of protecting those who suffer in this way (through no fault of their own and not through fecklessness) and of helping them to rebuild a new livelihood. But a welfare state can only be financed out of the fruits of success, so protection can be offered to a minority of sufferers only. Again, is such a system desirable?

The fourth issue is also linked to the cyclical nature of the free enterprise system. Its continued success depends on increasing efficiency, the ability to make more for less. Eventually this must mean that the market for any commodity becomes saturated; more can be made than can be sold, either because nobody wants any more or because nobody can afford any more. New commodities – including 'improved' variants of the original – must then be invented, to provide marketable products that will produce profits (rewards to free enterprise) and keep people employed – and buying. But what are those products? Do they represent recognition of needs which have previously not been met? Increasingly no. Many of them are weapons of destruction. Some are meant to be, as the escalating arms race shows. Other are not, but turn out to be, as with the new patent medicines that go wrong but are still foisted on a (deliberately maintained) ignorant market. Some are not products but services, bringing into the market place things (such as various forms of entertainment) that formerly people provided for themselves. And there are many other inventions, some of which bring renown to the inventors concerned, who must do new things in order to succeed professionally; many of those being produced today in the name of science and medicine are frightening, to say the least.

Fifthly, the free enterprise system is alienating, because of its treatment of the individual. Increased efficiency in production is almost invariably associated with routinization of labour; people must work like robots – or be replaced by them. Such alienation stimulates frustration, because the disciplines imposed suppress individuality and creative energy. Often this frustration is directed at the alienating instruments. The outcome may be relatively trivial – rude words in sticks of Blackpool rock, for example. But it may lead to greater unrest, and to a growth of crime and violence.

So is it really development? We measure that by very simple devices such as the level of GNP per capita; and we aggregate all expenditure together as if it were a contribution to increased welfare. Is more expenditure on police forces to suppress the unrest stimulated by alienation a contribution to welfare? Is more expenditure on arms to counter the competition for spheres of influence a contribution to welfare? Are we really measuring the quality of life (Johnston, 1976)?

Finally, it is not only the welfare of individuals that is at issue but also the welfare of the global ecosystem – even the future of the earth as a habitable planet. The drive to produce and consume more is a drive to exploit the earth's resource base more. Some parts of that base are irreplaceable; others can be replaced, but not at the same rate that we are removing them. And so as development proceeds the earth itself is being impoverished, if not raped.

Appropriate Development?

All of these issues suggest to me that the ideology of the free enterprise system, that development as defined within it as the best way forward for all, is inappropriate. And yet, we continue to act otherwise. For example, we suggest that disparities between places are soluble through policies that promote the interests of the less-advantaged; we usually call them regional development policies.

A little reflection quickly raises doubts, however. Let us take an example of a place with high levels of unemployment and of crime, with many problems of health, diet, and housing conditions, and much social unrest – Detroit, perhaps. Could we improve Detroit, provide jobs, reduce crime, improve health care, diet and housing conditions, and generate social harmony? Undoubtedly yes, but how would such improvement programmes be paid for? If the money were to be found within Detroit, then there would have to be some redistribution of wealth. Would the more affluent accept this? Or would they move, thereby in all probability exacerbating Detroit's problems? If the money were to be found elsewhere in the USA, then redistribution would be needed at another scale. Would the residents of other parts of America be prepared to make substantial subsidies to Detroit, especially since their home areas undoubtedly have problems too? In both cases, the potential is slight, but not because of any lack of goodwill. People in the rest of the USA would undoubtedly like to see everybody in Detroit prosperous – as long as their own prosperity did not suffer. And the rich and powerful of Detroit would like to see the city's problems gone – as long as the price were not substantial. But it can not happen.

Or can it? Can the USA as a whole not be prosperous, by being successful in the world? The free enterprise economists say that it can, by gaining the competitive edge over everywhere else – but the consequence is that the USA then imports profits and exports problems. Development is countered by underdevelopment, and everywhere wants to be developed. Detroit might prosper again, but if Detroit wins somewhere else loses. That conclusion is firmly based. What we do not know is how long a place can go on winning, while others go on losing.

Opting Out

The preceding paragraphs indicate a growing unease with the type of society in which I live. So why not either go to another society or opt out altogether? As some uninformed critics say to people who claim that things appear to be better arranged elsewhere: go and join them. So why don't I? And if I can find no society worth joining, or that will have me, why not just create my own?

With regard to the first part of that question, the answer is that I know of nowhere. There are countries where the free enterprise system does not reign, and where the organization of production, distribution and exchange does not follow the dictates of a market. The explicit goal is one of equality, of production and distribution to meet needs, but as yet they have not managed to produced much relative to perceived (by me) needs. The organization there is such that those in charge define their own needs differently and have structured society accordingly; they are implicitly producing inequalities too, and are doing it by directing human activity in ways that are as alienating as those of the free enterprise system. While I accept their ends, I cannot accept their means – which also involve the sacrifice of generations.

As to the second issue, the basic problem concerns access to the resources necessary for life. Land and water are necessities, in sufficient quantity and quality to support me and any who would join me. These can be obtained either by illegally occupying a piece of land, by purchasing it (presumably from the fruits of success in the free enterprise system), or by the good fortune of somebody giving it to me. All are possible, but unlikely; certainly there is little land available for squatting that could support many people.

Even were land available, could I survive if I opted out; would the system allow me to? As an individual I might, as long as I did not interfere with the system. But many individuals could not opt out without being a threat and they would soon be either forced out or priced out – as the history of the expansion of spheres of influence shows. Land is too precious to allow its sequestration. Furthermore, unless I were very fortunate – if I obtained a distant island – I could not stop the free enterprise system affecting me. I could not stop development up to the boundaries of my land, and the many externalities (most of them negative) that would spill over as a consequence; nor might I be able to prevent interruption of, for example, water supplies. Freedom to use my piece of land is constrained by the use made of others.

I may of course only want to opt out partially, to retain some links with the 'outside world'. In particular, I may want to obtain goods and services which would have to be paid for, presumably from the sale of the products of my own land and labour. Unless I have some monopoly, I will be competing with free enterprise producers, whose search for greater efficiency will increasingly undercut my price. I must produce and sell more to stand still, and sooner or later I will have to increase my efficiency. As the history of the European peasantry shows, sooner or later the free enterprise system takes over. So does the history of the small shop-keeper, someone whose desire for a satisfactory level of income only is continually threatened by the increased efficiency of supermarkets. And if I joined a cooperative the same would be true. To compete with non-cooperatives efficiency would have to be increased – and the competitors would be wanting to lure the successful cooperative manager away.

Of course, some do opt out in a whole variety of ways; by living off the welfare state, for example, or by becoming part of the black economy. But their degrees of freedom are limited by the success of that they have opted out of. Their numbers are limited too, and as soon as they pose any major threat they are removed by the policing powers of the free enterprise system.

ACHIEVING CHANGE THROUGH THE BALLOT BOX

The alternative to opting out would seem to be the achievement of change by democratic means, through the ballot box. I have participated in virtually every election – national, regional and local – for which I have been qualified to vote, and have been under the (implicit) impression, like most of my other fellow-voters, that by exercising my democratic choice I was taking part in the determination of what would be done for the public good in the territory concerned. (The fact that I have failed to pick the winner in virtually all of those elections is irrelevant.) And since about 1969 I have also been involved as an academic in the study of elections, analysing various aspects of who voted what, where; again, such study was initially based on the unconsidered belief that the operation of democracy involved people getting the sort of government and society that they wanted.

Or are they? The model of politics that I adopted – closely linked to Anthony Downs's (1957) pioneering work on public choice theory – was associated with the consumer sovereignty approach to economics that underpins the free enterprise model. In that approach, consumers decide what they want, express that through the market, and the suppliers of goods and services react accordingly. When applied to politics, voters decide what they want from governments and the contestants for government (political parties and their candidates) propose policies accordingly. Most voters, it seems, formulate their views on desirable policies according to their socio-economic position in society: thus the relatively affluent, with property and other investments, want governments whose policies would protect and promote those interests; whereas the relatively poor, with little or nothing to defend, want governments to redistribute income and wealth from rich to poor. A few (a majority in some societies, such as the USA) vote according to membership of other interest groups, such as religion and race, but economic self-interest is the major determinant.

This model provided the means of predicting – or rather, in my case, postdicting – how people would vote in different places. Not everybody conformed, of course. I didn't myself – or thought I didn't until I read David Robertson's (1984) book – but I classed myself as a deviant. Thus I could offer an explanation of the geography of how many voted what, where. The estimates were reasonably good, except that in most areas the level of support for the largest party was usually understated. Again, an explanation was at hand – and a geographical one at that. People discuss political issues with their neighbours, it said (again, I was a deviant!), so that in any area more supporters of the majority were likely to discuss with and convert supporters of the minority than vice versa; conversation leads to conversion. The explanation fitted the data so I accepted it, even though it didn't fit me; like the normal scientists in Kuhn's paradigm model, I preferred to ignore the anomalies.

This consumer sovereignty model of politics, set in a geographical context, assumes that voters decide what they want and parties set out to satisfy them. Thus the parties are relatively passive agents, conducting private polls, identifying the salient issues, and then producing a package of policies which will make them most attractive to a majority of the voters in a majority of the constituencies. But is this really so, any more than producers of goods for sale are passive responders to markets? Certainly there are policies that parties decide not to pursue because they believe they would be unacceptable to the electorate, but in general they structure the agenda, not respond to it. After all, politicians depend on electoral success every bit as much as businessmen depend on market success, and they will want to manipulate the agenda to their best advantage. Reflection on the choices available showed that the menu is being drawn up not by me and my fellow-voters but by the parties. As Schattschneider (1960) describes it in his classic *The Semi-Sovereign People*, they are biasing the menu – indeed 'all organization is bias'. They stress some issues and avoid others, and ask for voter confidence in them as people rather than in what they stand for.

But this is a democracy, and so if the package that I want is not on the menu, then if I feel strongly enough about it I can place it on the menu and discover whether others agree with me. People do this occasionally, almost invariably unsuccessfully, which suggests that the consumer sovereignty model is right and that people are being offered a reasonable choice. But the problem is not necessarily one of relevance; rather it is a question of access. To be successful a political candidate/party has to be 'sold' to the voters; their support must be 'bought' by the party convincing them both that its policies are sensible and that its leaders are trustworthy. In a mass democracy, this can only be achieved through exposure to a large number of people, which can usually only be bought, either directly (as with the immense expenditure on TV and other advertising in US Presidential campaigns) or indirectly (by getting the support of powerful elements in the media, especially the press). Thus people with money have to be won over to provide sponsorship, not just for one pre-election campaign but for continuous promotional activity that will keep the party, its personalities and its policies in the public eye.

Where can such sponsorship be obtained? Occasionally, altruistic, very rich people will provide it, but not for continued activity. Long-term, large-scale support depends on building an organization around vested interest groups, who believe that (indeed are often promised that) they will reap a substantial benefit from their investment when the party is returned to power. In Britain there are two major such groups: capital and the trades unions. No other group, to date, has been able to mobilize long-term support, and for more than 50 years now the two parties associated with them (Conservative and Labour respectively) have dominated British politics. Until the early 1980s, the only major threat came from parties with a particularistic, and therefore minority, base, notably the Scottish, Ulster and Welsh nationalists. The Liberal Party has attempted to build on local, or community, politics and in alliance with the SDP has promoted a non-vested interest (i.e. national) appeal; as yet, however, it has failed to find either massive continual sponsorship or sufficient votes (in sufficient places) in the contests that matter – general elections.

So the choice is limited, essentially to two policy platforms. But is it really a major choice, between a pro-capital or a pro-workers party? The former certainly promotes policies that favour capital (i.e. free enterprise), whereas the latter advocates redistribution and state control. But both are constrained. For the Conservative Party, advancing the goal of profit-making to the detriment of anything else is likely to alienate workers and stimulate social unrest, so that unless it is prepared for coercive policies the party must temper its advocacy; it does, of course, seek to counter potential unrest by the use of ideology, arguing that its pro-free-enterprise policies are in the best long-term interests of everybody. Similarly, Labour cannot push redistribution and state control too far, or it will discourage investment, stimulate unemployment and fiscal problems, and generate its own problems. Ultimately, then, both must promote free enterprise yet retain worker support; they offer variations on a single theme.

But it is argued that those variations have become greater in recent years, with the one party becoming more stridently radical on the right and the other more so on the left. The result, according to the adherents of the 'adversary politics' school (Finer, 1975; Gamble and Walkland, 1984), is uncertainty in the country; both parties are scaring off investment because of the difficulties of predicting what government policy will be after the next election. What we need, some argue, is stability. This will be achieved by electoral reform because the electorate are really middle-of-the-road, and if they had an electoral system that allowed them to express this they would give a permanent centrist government; more choice would mean less choice.

So do I really have choice, or am I really selecting between slightly different ways of running society along present lines – i.e. the domination of free enterprise? The latter, it seems. The sort of democracy that I live under – liberal democracy – allows me choice of how I want the free enterprise system to be run, but not of whether I want to replace it. This is quite simply because those who advocate the latter could not survive long enough to see their changes through. They could not get sponsors, they could not disseminate information, and even if they gain some electoral success (as with the Poujadistes in France, Glistrup in Denmark, Allende in Chile, and Manley in Jamaica) they would soon find their ability to survive within the economic system immensely constrained; even mildly reformist governments like those of Labour in Britain have been rapidly disciplined by the forces of free enterprise and its allies, such as the IMF.

Of course, the choice that I have is apparently greater than that available to my contemporaries in the Second and Third Worlds, and mild dissent is not crushed as ruthlessly as it has been in Czechoslavakia, Hungary, Poland and Zimbabwe, for example. There, dictatorship is the norm, almost invariably with a strong military backing. The justification for this is usually that it is necessary to promote the stability that is crucial to economic development. Only the First World can offer real choice, it seems. But that choice is a consequence of development that is built on underdevelopment elsewhere. So can there ever be choice everywhere?

All of this discussion has been set at the level of national government. But it applies just as well to local government, as is crystal clear to a resident of Sheffield. Some have choice; some have more than others. But we are locked into an economic system that must be fostered through the political system. Slight variations are possible, with local nuances, but no more. Choice in the political market place, as in the economic market place, is very constrained, and there is little that we can do to break down those constraints.

RECREATION AND LEISURE

The conclusions that I draw from the previous sections is that there is little I can do to influence the economic system. So perhaps I should make the most of it, and enjoy the rewards. In particular, I should make the most of my leisure time and the opportunities for recreation. Then, it is argued, I am free to choose – to pursue hobbies, to travel, to read, to watch sport and TV etc.

Once again, the superficial appearance is one of choice; consumer sovereignty reigns. My major leisure pursuit is campanology, change-ringing on church bells. This I do, with others, on bells provided and maintained by church authorities for their own purposes but which we are free – within constraints – to use for ours; we have entered an unwritten contract with the church authorities that if we agree to ring the bells when they wish they will allow us to indulge our interests at other times. Nobody makes me do this; I make myself.

Just as nobody requires me to ring bells, so nobody requires me to read during my leisure time, let alone requires me to read certain books; the choice is entirely mine. And the same is true of the programmes I watch on TV, the time and place of my holidays, and so on. In my leisure at least, it seems, I choose.

But all of my choices are conditional ones, dependent at least in part on the decisions and actions of others. If sponsors were not prepared to donate towers and bells to churches, then I could not ring; I certainly could not afford, with a few others, to provide and maintain my own and the most I could do would be encourage the church authorities to raise money for bells. I have some influence there, but not a lot. On other aspects of my leisure time, I have even less. If investors are not prepared to put money into the creation of certain types of TV programme, the publication of certain types of book, or the construction of hotels in particular places, then my consumption opportunities are limited. And there is little that I can do individually to change the situation.

The constraints on my choices are largely, though not entirely, materialist because the decision to provide a certain leisure service is usually a financial one; in the spirit of the free-enterprise system the basic question being asked is 'can a large enough market be stimulated for this service so that the profit will be larger than that obtainable from other channels of investment?' Many TV programmes are made either to sell or to attract advertising. Most books are published by commercial enterprises etc. Most 'news'-papers survive on the sale of advertising space. So are the programme-makers, the newspaper and book publishers, the hotel-builders and many others merely reacting to tastes; or do they create them? What are tastes; innate needs or created desires? Is it natural for children to want to build sandcastles on beaches, for adults to watch soap operas, to be stimulated by certain types of picture and performance, to appreciate mountainous landscapes and to want a sun tan, or are they in some way conditioned into thinking that such things are desirable?

Attitudes and values are not innate, they are learned. A major purpose of education systems is to inculcate certain attitudes and values – those which are considered desirable in the socio-economic context. And in a materialist society, many of those attitudes relate to goods and services that can only be bought. Anything that is bought is also sold, and in order to be successful the sellers must mould our attitudes so that we are positively inclined towards their wares. This is the function of the advertising industry. It is, of course, backed by market research, which is designed not to identify our innate needs but to see if, given our lifelong conditioning, we are likely to react favourably. Sometimes the results are not very good because a faulty decision is made – remember ten-pin bowling. We can choose not to patronize a particular good or service – especially as not to buy any is a viable option since leisure consumption is not necessary for survival. But we can only choose from what is provided.

In recent decades, leisure has become a big business, as it has offered greater potential profits than many other areas of production. But this has not necessarily increased choice. Indeed, in some areas – professional sport, for example – it has narrowed it. The market forces of free enterprise are now major arbiters of taste in a wide range of leisure and recreation activities, constraining rather than opening up choice. Of course we can opt out and engage in non-commercialized leisure activities – gardening (though this is increasingly commercialized); amateur dramatics; barber-shop quartet singing – but if we opt in, we opt for what the market thinks it should provide, and we should consume.

LOOKING AROUND

The conclusion that I might draw from the above observations is of a world with little variety in it, dominated by the free enterprise system. And in part it is, as the market forces seek to create 'universal cultures' – I want to buy the world a coke. But yet, as those who seek to sell me holidays make clear, it is also not like that – and my own travels confirm it. The world is an immensely complex mosaic of different places, not only in their physical environments but also in their social, cultural and political milieux. The same economic forces are apparently producing different outcomes. Why? Common sense and casual observation tell me that it is not environmental determinism, and my research activities in Australia and New Zealand as well as Britain have taught me to beware the wholesale importing of North American ideas.

And so I conclude that although I am materially very comfortable I am unhappy with the world that I live in. I want to change it, so that everybody is materially comfortable and has a high life expectancy. But I do not want this to be bought at the expense of freedom. I want choice and fulfilment, for all. I accept that I have choice; however, it is constrained not by me but by forces apparently beyond my control.

In order to conquer those forces, I need to understand them more fully because only through such understanding can I hope to build a new future with others. Understanding, and communicating that understanding, are vital. Such a task is extremely large, as I realize when I face the complexity of the world. The remainder of this book outlines the path I am following, as a human geographer.

3

A Model of Society

We see what institutions allow us to see, we know what institutions allow us to know, and we act as institutions allow us to act.

Stephen Gale, 1982, p. 61

The previous chapter has outlined a series of observations relating to contemporary society, drawing on both academic and general perspectives. They present neither a coherent nor a complete critique of that society, nor are they intended to. What they do provide is a framework within which my approach to human geography is evolving. In particular they have led to to the following conclusions.

1   However accurately I am able to portray the empirical world, this does not provide me with an understanding of why it is as it is, for two reasons: first, the terminology with which I make my portrayal is one that I learn in a particular context, so it is not neutral; and secondly, though the portrayal may tell me what the environment is in which I make choices it tells me neither why that environment contains what it does nor why I have been predisposed to describe it in that particular way.

2   Following from this, my decision-making takes place in a social context which comprises certain imperatives but also a great deal of freedom. At their most basic, the imperatives are those of survival – of obtaining food, water, shelter etc. – and reproduction. In addition, and most importantly, the means of fulfilling those imperatives is structured for me. In the society of which I am a part, food, drink, shelter and clothing are all commodities available to me only through purchase (or as a gift from somebody else who has purchased them), and the sole way in which I can make those purchases is to sell my labour and obtain the necessary money. Further, there are great pressures on me to purchase much more than I need to survive, both more food and drink than are necessary for survival and a great range of other goods and services that are available only as commodities.

Although I must follow these imperatives in a general sense, there is no necessity for me to do so in any particular way. My behaviour is not so conditioned that my actions can be predicted with absolute certainty. To whom I sell my labour and for what and the particular goods and services that I buy are all the focus of choices – albeit constrained choices because only certain options are available to me and I have little influence over the supply of goods and services. Like everybody else, I must make a series of choices, though the nature and range of the options available to us may vary.

3   The empirical world is the outcome of the exercise of choices. The pattern of land use in rural areas reflects choices of how to use the physical environment, for example, whereas the patterns within the built environment represent choices of where to do certain things in an urban context. The institutions of society are also the outcome of choices regarding how it should be organized, as are the ruling ideas into which we are socialized, at home and through the education system in particular. All of these are the context for further choices; the results of others' choices enable me to choose, but also constrain me to certain options.

By making choices, I contribute to both the maintenance of and change within the empirical world, thereby contributing to the environment for future choices, mine and others'. By choosing a particular career and employer, a home and so on I am making my future more certain, and that of those dependent upon me for survival. I am not closing down choices, for I could break away, although that is not very probable. Those dependent upon me could repudiate my choices too, as they set out to maintain themselves. Thus I influence others as well as myself, and many are influenced indirectly by my choices: if I take a certain job and rent a certain home, then those options are closed to others. Few if any of my choices will have a major impact on society, but I am actively involved in its reproduction in a particular format which can have a minor effect on the lives of many.

4   The context within which I act, and which I am involved in reproducing, is profoundly geographical. My options are structured within a local environment – physical, built, political, institutional and cultural – parts of which (notably the political, some aspects of the built, and a few elements of the physical) are territorially demarcated and many others of which are spatially restricted if not circumscribed. I am not a prisoner of that environment, but I am strongly tied to it for it has socialized me into a certain set of attitudes shared by others in that locale. I could escape both

physically (if another society would accept me) and mentally, if I used the modes of thinking made available to me in particular ways. Basically I am geographically constrained: my ways of thinking, reasoning, communicating, surviving etc. have been learned in an environmental context, and were I to move to another context – should the option be available – some of those ways would have to change.

This spatial constraint is largely one of history. The culture that I have been socialized into developed as a local response to a particular physical environment, with basic means of surviving and the associated social organization being put in place, and then reproduced, by the people who chose to live there. As successive generations reproduced that society, so they changed it, in response to the dynamics of the society itself and their contacts with members of other societies, whose responses may have been somewhat different. In particular, many societies have expanded and come into contact and conflict with others; the accommodations that they have reached have been part of the development of their milieux. Thus there are cultural geographies, varying interpretations of how societies should be organized in order to survive.

Three Propositions

These general conclusions lead me to the belief that to be viable the study of human geography must be firmly based in an understanding of societies, how they are organized and operate. The nature of that understanding is to be founded on the following three propositions.

The crucial element is the human actor. Individuals have the powers of rational thought, which they operate in a variety of ways according to the materials made available to them in their cultural contexts. The powers must be developed, therefore, We are born with the abilities to assimilate and use information, but not with information; we must foster those abilities. How we do so depends upon who we are in contact with, for they provide the information and they teach us how to use it, how to reason and express ourselves. The world is the result of human thinking, which has taken place in contexts that are both enabling and constraining.

In order to survive, humans have created basic *organizational forms within which their lives are structured*. Those structures are neither fixed nor necessities (with one exception, as outlined below); we live within them and reproduce them, but we could also change them or even replace them.

The interactions of human agents and the structures that they have created lead to *the formation of societies, complex empirical organizational frameworks.* These comprise a vast array of formal and informal protocols, which are deeply etched into our ways of thinking – through socialization – and into the landscape within which these societies operate.

What we have, then, is a system of three interacting elements: people, mechanisms and societies (figure 3.1). People are central to this system, because they occupy the intermediate position; mechanisms and societies only interact through people. Mechanisms are created by people as interpretations of the basic need to survive, and societies are then created as interpretations of those mechanisms, themselves bound together in structures. From then on, people interact with societies as they live out their daily lives. They interpret and change societies; and they interpret and change mechanisms.

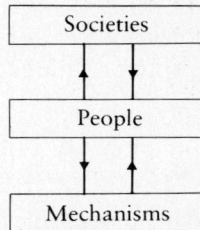

Figure 3.1 *People mediate between societies and mechanisms, both of which are created and reproduced by them*

As human geographers, our basic interests are in understanding certain aspects of the empirical world, of the societies and the people in them. But to understand those people and their societies, the worlds that they have made, the worlds that they live in, and the worlds that they are making, we must understand the mechanisms, because they provide the essential framework for everything else. This is the goal of the remainder of the present chapter. In it, I do not deal with a range of mechanisms but with only one – the capitalist mode of production – because it dominates the world in which we live at the present time. Nor do I present a full and rounded discussion of capitalism (in part because other geographers, notably Harvey (1982), have done it much better than I could). For my purposes, an outline sketch is sufficient.

MODE OF PRODUCTION

The capitalist mode of production, usually referred to as capitalism, is a structure evolved by human decision-makers as a way of enabling people to fulfil their material needs. In many respects it has been remarkably successful in this, as indicated by the major increases in, for example, life expectancy in recent centuries – although not all societies have experienced similar average increases. But it is a structure with inherent contradictions, the resolution of which is the cause of conflict.

Capitalism is a materialist structure, which treats the means of production – land and labour – as commodities to be bought and sold. The necessities of life are produced neither collectively nor cooperatively, but in a context where the raw materials (land) and the ability to work them (labour) are bought in the requisite quantities and put to work to produce the desired output. The products are then sold, being purchased with the wages received for the sale of labour power. Central to the organization of both production and exchange is money, which is used as a universal currency in the buying and selling of materials, labour and products.

Money is a social creation, not a natural commodity. Within capitalism it is a necessity, for nothing can be bought without it (or without the promise that it will be handed over by a specified date after the purchase is made). Therefore, possession of money is a major source of power within capitalist society. But money can be used in two ways: (1) to purchase commodities; and (2) to purchase materials and labour for the production and sale of commodities. Power is associated with both. Those with more money are better able to buy commodities and so influence the operations of markets: they have purchasing power. But they are also better able to create products, and by doing so create employment without which some people may have no purchasing power. Those with enough money to use some of it in the latter way have structural power.

Capitalism developed on the basis of these two uses of money, and involved the separation of the population into two groups. The first, and most numerous, were those who used money only for the purchase of commodities. The second, a minority, used some of their money to purchase commodities but much of it to produce commodities, and thereby create more money. They were able to do this for one of a variety of reasons, relating to the structure of pre-capitalist society, including: unequal ownership of land resources; control over labour, through a variety of practices (e.g. slavery, serfdom); and the accumulation of money as the profits from trade. The structural power that they possessed enabled them to gain control over land resources, and thus to influence very strongly (if not dictate) the organization of society. The majority had no means of survival other than their labour power, which they had to sell to those who used their money to create products.

The rationales for investing in production were the enhanced purchasing power that it created, which could be used in conspicuous consumption and the display of wealth, plus the ability to accumulate wealth through a continuous cyclical process of investment and return. That process is denoted by the simple sequence:

$$M - C - M^1$$

in which money (M) is used to produce a commodity (C), which returns money (M^1) in greater volume than the original investment (i.e. $M^1 > M$). Only the guarantee, or virtually so, that the return will be greater than the investment will ensure that the latter occurs: as Harvey (1982, p. 13) expresses it: 'The only possible motivation for putting money into circulation on a repeated basis is to obtain more of it at the end than was possessed at the beginning.' What is produced (C) is irrelevant, so long as it can be sold at a profit; its utility is secondary to its saleability (indeed saleability is the definition of utility).

Capitalist society was built on the existence of two major groups, those who advance money (capital) in order to make more of it, and those who work for money in order to purchase the means of subsistence – with subsistence defined socially (what it is considered desirable to purchase) rather than absolutely (what is necessary for survival and reproduction). People in the first category are also in the second, but not vice versa. These two groups became known as classes – capitalist, or bourgeoisie, and proletariat respectively – and because their existence and inter-relationships were fundamental aspects of the structure, a capitalist society is identified as a class society. The classes are interdependent, because in the long term neither can survive without the other; the capitalists need the proletariat to work and to buy; the proletariat need the capitalists to invest. This interdependence is not one of equals, however, because in the short term the capitalists can survive on their accumulated wealth even if they have no investments earning income, but if they have no paid work the proletariat cannot consume. Economic power in the structure is thus unequal; wealth and power are strongly, and positively, correlated.

A capitalist class society is one with an inbuilt conflict between the two groups, fundamentally because of the need for M^1 to be greater than M. Accumulation of wealth could be seen as a function of greed, so that if capitalist avarice were tempered the drive for more and more money would be lessened. But this does not allow for the constraints of the mode of production itself. Capitalists themselves are competing to sell their products, and on the assumption that there is no qualitative difference between them, the cheapest products should capture the market, or at least a major share of it. Thus a capitalist whose price is relatively high will be unable to sell all that is produced, at least at a price that yields a profit. As a result, instead of accumulating wealth that capitalist will lose it. To survive in a free market – that is, to make an acceptable profit – every capitalist must seek to undercut the price of all others, or face economic disaster.

The need to compete and the desire to accumulate as much as possible are linked in the search for high levels of profitability. To create a product, money must be invested in the materials to be used, the labour to be applied, the resources (buildings, machinery etc.) involved in the transformation of the materials into the product, and the labour involved in its sale. These are grouped into two main categories; labour, or variable capital (v), and constant capital (c). (Constant capital is the output of earlier rounds of investment, perhaps by others, in the preparation of materials and the construction of buildings etc., and so is also known as 'dead labour'.) The return from the investment is the price of the product (r), and the surplus value produced (s) is the difference between that price and the costs of production: $s = r - (c + v)$. The rate of profit (p) is the ratio of the return to the outlay: $p = s/(c + v)$ or, since $M = (c + v)$ and $M^1 = r$, $p = (M^1 - M)/M$.

In a competitive situation, the capitalist aims to keep r at least as low as the price charged by any other producer, and also to keep p as high as possible. The size of p is a function of the size of s relative to either or both of c and v, given that in a market with many sellers no one capitalist can manipulate all values of r. Thus only c and v can be manipulated. In any particular production process, the costs of c can be separated from those of v, but in the total production process (that is, the production of the machinery used as well as the final product) c is itself a function of v, since constant capital is a product of labour at an earlier stage in the process. Ultimately, therefore, only the variable costs of labour (v) are manipulable; their size is fundamental to the rate of profit, if r is fixed.

To increase profitability over time, or even to remain profitable in the face of competitors, capitalists must reduce their labour costs. They can do this in a variety of ways. They could simply reduce the wages paid to labour, but this is a strategy which in the short term produces worker dissent (it reduces their immediate purchasing power and therefore produces a decline in their standard of living, if not their ability to reproduce), which could disrupt production and reduce sales and total profits: in the longer term the decline in purchasing power would limit the ability to sell all that is produced, and again eat into profitability. Or they could urge workers to produce more, to work harder, so that their gross wage might increase (as a reward for greater productivity) whereas the amount paid per unit of production would decrease. Or they could reduce the labour cost (v) relative to constant costs (c) – thereby increasing what is known as the organic composition of capital: $c/(c + v)$ – on the argument that the dead labour in machinery is more productive than is live labour, and less skilled, cheaper labour could be used instead of expensive, skilled labour to operate it. Whichever strategy, or combination of strategies, is used, however, labour loses.

Ultimately, both classes will lose. Any strategy that involves an increase of productivity will mean a rise in production. The market will be flooded with more goods. But who will buy them, for the relative purchasing power of the workers may have declined? Efforts can be made to expand the market in a variety of ways, to encourage greater consumption, but eventually the market will be saturated; more will be available than people want, or are able, to consume. To maintain their incomes, the capitalists compete for a larger share of a shrinking market, by cutting prices, urging greater productivity etc. Eventually, however, the only solution is to produce less, which means employ less labour.

There is, therefore, continual conflict between capitalists and proletariat. The former urge greater productivity, seeking always to reduce how much they pay to labour, and the proletariat for their part see their standard of living under constant threat. The latter may yield, because they see no option, agreeing to accept new machines, less skilled tasks, lower relative, even absolute, wages etc. in order to stay in work. But eventually they are told there is no work, because there is no market for what they produce.

And yet there is no apparent alternative. The capitalists could give all of the surplus (s) back to the workers, in higher wages. This would increase their purchasing power, but because there is no investment in production there is no more for them to buy. Their increased purchasing power would soon be eroded by inflation, and the lack of new products would soon lead to a decline in employment. Similarly, the capitalists could spend it all on consumption, but again this would be a short boom only, followed by a slump generated by the absence of anything to buy. Or they could just save it, with the same consequence. Only if they reinvest it, or its major portion, in more production, for more profits, will jobs be maintained and reproduction assured.

But, as just indicated, continued investment in production is ultimately self-defeating, because a saturated market means that selling becomes increasingly difficult; there is a crisis of overproduction, or underconsumption. The only sensible reaction to this is to redirect investment into something that will sell, which could well mean writing-off much of the previous investment in constant capital, dismissing the labour force, and starting again. What in? New products must be 'invented' and markets stimulated, a procedure that should be continuous and should not wait until the boom in a particular product ends. Hence investment in research and development, and in advertising and marketing, in the continual search for products (goods and services – see the section on leisure in the previous chapter) that people can be convinced they need to buy.

While this restructuring process continues, the competition to sell established products leads to changes in the organization of production. Successful producers/marketers will either force others out of business, thereby capturing a larger share of the market, or will be able to buy out weaker competitors, with the same consequence. Thus production becomes concentrated in fewer, larger units. And the search to diversify means that those large, successful units spread their interests into other areas of production too, so that concentration in one sector of the economy is followed by centralization as the major units gain large market shares in several sectors.

Concentration and centralization have been major trends in the evolution of capitalism. It could be simply that the assets of successful individual capitalists expand, and wealth becomes highly concentrated. Alongside this, however, there may be a diffusion of wealth. The scale of investment needed in the research and development of new products and the establishment of production lines may be beyond the ability of one investor or a few; even if it is not, they may not wish to commit too much to the scheme, in case it fails. They want to spread the risk, and to raise investment capital from others. In this way, capitalists combine, in a variety of ways, submerging their individual identities in a corporate unit. And they may encourage members of the proletariat to lend them money, directly or indirectly, with the promise of a share of the profits.

With the growth of the large corporations and the search for more investment from a wider range of sources, individual capitalists decline in importance, and are replaced by financiers and financial institutions whose role it is to amalgamate, invest, and manage money. Ownership and management become increasingly divorced, but the goal is the same: creating profits for the investors. If these are not seen to be forthcoming, investment slows down, production declines, and jobs are lost. A system created by individuals has been, at least partially, taken over by institutions but the dynamic remains the same. The institutions must continue to accumulate, if they are not to collapse.

A major consequence of the growth of these institutions in all areas of capital – industrial, merchant and financial – is an alteration in the relations of production and in the class system. The simple dichotomy between capitalists and proletariat becomes an oversimplification. (Some claim that it always was, that there were always many people who could not readily be located in one of the two classes. To some extent this was so, but more importantly it reflected the survival of pre-capitalist forms – small shopkeepers, peasant farmers, craftsmen etc. – which had not yet been destroyed by competition with capitalist producers and distributors.) There is the growth of what we know as the middle classes, those who operate the expanded capitalism for a salary. In some ways they are wage-labourers like the proletariat, but they differ from them in their responsibilities. Their range is wide, because they are responsible for attracting and amalgamating investment money, deciding how to invest it, and managing it: many of their tasks (accountancy, parts of the legal services, management etc.) have been professionalized, again setting them apart from the proletariat. Their rewards are generally high – in part, at least, because they (as a group if not individually) decide on the levels to be paid, and in part because they are paid on results; the greater the profits, the greater their salaries, making them very like the capitalists they have replaced. Though people with substantial wealth derived from operation of the capitalist:proletariat relations of production remain, few of them continue as owner-managers. Most invest their wealth through the institutions, obtaining powerful positions in them as a consequence in many cases, and hand over the day-to-day running of the system to the new managerial class; they still draw their profits, and will shift their money if the rate of return slackens.

To some extent, therefore, the dynamic of capitalist accumulation sees the original capitalist class required to share its economic power with the new professional-middle class, as the best (possibly only) way forward in an increasingly complex procedure of accumulation based on ever-larger conglomerates. The employers of labour are increasingly the managers of other people's money, rather than of their own. They are still seekers of profits, for if they do not produce a return for the people whose money they are handling they will probably lose their positions. (Some claim that, because it is other people's money they are handling, the managers are less prepared to take high risks than were their capitalist predecessors. They are more conservative in their business strategies, more concerned with satisfactory market shares and profit levels than with the gambles that could bring either even higher profits, or failure.)

This change in the structure of one class in the relations of production means that economic power in the capitalist system is somewhat more widely dispersed, though not much. The maintenance of the capitalist dynamic requires a managerial class (with many levels within it), but one which is firmly committed to the goal of accumulation. (Again, it is claimed by some that the policy of encouraging proletariat investment in the system further removes the class differential; just as the decline of the individual capitalists means that 'we are all workers now' so the spread of investment, both individually and through institutions like pension funds, means that 'we are all capitalists now'. But these 'petty investors' have little, if any, economic power. Individually they can do little to influence the use of their money and the rules drawn up for the institutions by the professionals – such as for trade union pension funds – ensure that use of their funds is entirely consistent with the managerial norms.) The class norms are certainly blurred and socio-economic mobility – into and out of the managerial class – increases, but the majority of the proletariat remain wage-labourers.

To counter this relative powerlessness, wage-labourers may seek to contest the powerful forces – the capitalists and the managers – collectively rather than individually. They have done this through the trade union movement, by developing bodies whose sole purpose is to advance the interests of their members through increased real wages, better working conditions and terms of employment, and so on. The promotion of collective bargaining is difficult when there are many employers concerned, but the growing centralization and concentration of capital -- with the consequent growth in the labour force of individual firms – increases the possibility of workers' solidarity and makes their collective power a greater threat to that of the employers. As such, they may be able to win concessions both in direct confrontation with employers and indirectly (through the state – see below), at least in the short term, because of the threat they pose to profitability if they withdraw labour and prevent others being taken on in their place. But that power is generally transitory, for two main reasons. First, the proletariat generally have little in the way of accumulated resources and are less able than either the capitalists or the large financial institutions to sit out a strike; the need to reproduce eventually forces them back to work, perhaps with concessions to the employers. Secondly, if they win concessions these can be self-defeating unless increased productivity is part of the bargain, because those concessions can destroy the profitability of the investment that keeps them in work. (It is for these reasons that workers seek concessions indirectly via the state. If the state imposes certain conditions of employment on all employers, or minimum wage levels in certain industries, then all are bound by them and no individual firm's profitability is more threatened. But, as detailed below, the state can only go so far in allowing such concessions because its power extends over a defined territory only. Lesser concessions in other states mean that the problem remains.)

Concentration and centralization of economic power, combined with diffusion of ownership, are major strategies that have developed in the face of the inevitable slumps that follow market saturation for a certain product. They are part of the restructuring which takes place continuously, as capitalism is reorganized in a great range of ways to try and ensure profitability. Such restructuring needs to be based on confidence, however. Capital will not be invested in ventures, new or old, if the prospect of profitable returns is small, whether because of market saturation, the uncertainties of new markets, problems with a recalcitrant labour force, or a range of other reasons. In such conditions, the rate of investment will slow down, not just in one sector of the system but in many; people with money to invest are wary of committing it to risky projects. The consequence is a general slump – low levels of investment, a reduction of employment prospects, a decline in purchasing power, and so on – all of which stimulate a spiral of decline. Many investments – in constant capital (buildings, machinery etc.) and in variable capital (for example, particular labour force skills) – must be written off, and bankruptcies become common. Some producers survive, and their market share increases. If confidence can be increased, then they are well placed to benefit from a new round of investment. Meanwhile, however, the system creaks, and many individuals within it suffer, in the managerial sector as well as the proletariat.

THE STATE

It could be implied from the description of capitalism presented so far that it is a self-regulating system. This is far from the case. The individual capitalists and their corporate successors are involved in competition, which they all want to win in order to increase their profits and wealth. Unless there are rules for that competition, the system is likely to be self-destructing.

To take one example: much of the buying and selling of materials, goods, services and labour involves the making of contracts that contain promises to pay certain prices, to deliver defined quantities and qualities at specified times, and so on. But what guarantees are there that the contract terms will be met? None, unless there are rules that make contracts binding, and there are agencies which ensure that the terms are met and contracts honoured.

The honouring of contracts would appear to be in everybody's interests, since it provides an environment of certainty within which bargains can be made and promises accepted. But unless there are certain sanctions for failing to honour contracts, it is not necessarily in any one person's interests to do so. If you can get away with not honouring a contract, and can be better-off as a consequence, then why not do it. The long-term consequence may be that others follow you and the system collapses because of the anarchic uncertainty, but you are only interested in the short term, in today's profits.

This situation is one example of what is known in the mathematics of game theory as the prisoner's dilemma. The best solution, in everybody's interests, is cooperation − in this case, honouring contracts. But no individual will agree to cooperate unless all others do, because greater benefits could be gained by stepping out of line. Indeed, unless cooperation is binding, it is in nobody's interests to agree; they can benefit by being outside the agreement. Thus nobody will agree, and the optimum solution is not attained. To achieve the necessary cooperation, it must be insisted upon by an outside institution whose power to make and enforce rules is accepted.

The enforcement of contracts is one element only in the environment that is necessary for capitalist cooperation and success. It provides one element of certainty, one part of the general guarantee that investors are looking for. There are many other such elements. People entering contracts accept promises that they will receive given sums of money on specified dates. They will want some assurance that on those dates the stated sums of money will be worth at least what they are worth now, that is, have the same purchasing power. The individuals with whom the contracts are being signed cannot offer that assurance, because they have no control over the value of money and the various factors − the money supply etc. − which influence it. Again, therefore, the environment is uncertain. To reduce that uncertainty, an institution which controls the value of money is needed, otherwise the stimulus to invest is reduced.

The discussion so far in this section has dealt almost entirely with relationships within the capitalist class and the need for a body to regulate and oversee them. Similar arguments can be raised regarding the relationships between the capitalist class and the proletariat. These, as indicated above, are conflict-riven, because the profit-seeking goal of capital necessarily involves, in the short term, strategies intended to reduce the real level of wages paid to employees and, in the long term, restructuring which destroys employment – albeit perhaps to replace it with other employment, not necessarily on such good terms. The proletariat, individually and collectively, counter these strategies. Their major bargaining power is short-term only, because it involves their withdrawal of labour and the destruction of profits which, of course, destroys their earning capacity too. But such short-term threats to profitability can be sufficient to deter investors. Unless they have a compliant labour force, they will not invest, with deleterious consequences for all.

How can a compliant labour force be achieved? In part, through concessions – higher wages, better working conditions etc. – but these are only feasible for profitable units, and even so in the long term are likely to reduce the return to capital, unless greater productivity is yielded in return. The employers could use repressive means, but these are likely to be counter-productive too. Workforce compliance must be guaranteed by an outside body, one whose power and authority is accepted by both sides. There must be rules for employer-employee relationships, which are accepted and enforced; if they are not enforced, then it is in no one employer's interests to operate them. Inter-class tensions must be contained by a neutral arbitrator, which ensures worker compliance for the capitalists and reasonable working conditions for the proletariat.

In both sets of relationships it has been shown that the continuity of capitalism requires an independent body or institution which regulates competition and mediates inter-class conflict. That institution is what we know as the state, a necessary body for the successful operation of the capitalist mode of production. It undertakes three basic roles.

1 First, it acts as the *promoter of accumulation*, by providing an environment in which investment takes place; it cannot guarantee profits, but can create the situation in which investors are prepared to gamble on them.
2 Secondly, it acts as the *legitimator of capitalism*, by ensuring proletarian acceptance of what is, at root, a system of economic relationships based on their exploitation.
3 Thirdly, it acts as the *creator of social consensus and order*, welding the conflicting parties together into an integrated society, in which the other two roles are possible.

To perform these roles, it undertakes a very large number of tasks, and is itself a substantial body divided into many state apparatus and sub-apparatus (as described by Clark and Dear, 1984). It is empirically separate from the classes within society, being the 'possession' of neither, in order to maintain its apparent neutrality. Its goal, however, is to promote the continuation of the society, and since capitalist society is built on inequalities (of wealth and of power) then in its actions it sustains those inequalities and the system that produces them.

The state is rule-setter and -enforcer, therefore, and is accepted because it, through the concept of sovereignty, is the sole repository of certain types of power within society, especially those associated with violence (military and police). Those are reserve powers, however, and the state prefers to work by promoting consensus, because this is both less costly and most likely to be conducive to investment and productive work. It acts by seeking to meet the demands of all, contradictory though this may seem. Thus, for example, as employers invest in research and development of new products they want a trained workforce to do the research, to pioneer the production, to operate the new plant, and to market the output. To undertake all the training themselves would be expensive, and those trained may then move off to other employers and the expertise is lost. Much more satisfactory would be the ability to draw on a trained labour force with the requisite general skills that could be quickly and cheaply honed to the particular task. And so the state provides an education system, paid for out of its taxation revenues so that all are contributing to the costs. In that way it promotes accumulation, especially if it favours the development of certain types of skill. But it can also use the provision of education in its legitimation activities. If employment, especially highly paid employment, is attainable through education, and the education system is open to all, then the state can claim that it is providing opportunities for the proletariat; their success is being subsidized too. Both classes benefit in aggregate, though the distribution of benefits is likely to be more uneven within the proletariat. To ensure that the unevenness is not itself a source of conflict, the state must promote an ideology of 'equal opportunity' that leads to proletarian acceptance of the idea that the competition for the best jobs is fair.

If the state fails in its attempts to promote accumulation, then it generates an accumulation crisis with part, if not all, of the capitalist class dissatisfied with its policies; changes are called for which will lead to renewed accumulation. If it fails in its attempts to legitimate capitalism to the proletariat, there is a legitimation crisis, with calls for new policies that will satisfy the demands of labour. If the two overlap, it faces a rationality crisis;

it is satisfying neither, and the social consensus is under threat. A rationality crisis is most likely to occur in a period of general slump, when investment is very low and unemployment high. Both classes are calling for new policies, and what they want is likely to be contradictory; the state must seek to satisfy both (or all, since neither class is united and there are many separate interest groups within each).

Each crisis is different, and calls for a new response, because under capitalism history does not repeat itself. As already indicated, capitalism is a continuous process of restructuring, of creating new situations out of the contradictions of the old. The failure of many small firms in one slump may lead to the creation of a few large ones, fostered by relevant state policies. But these policies will be largely irrelevant in the next slump, when it is large firms that are failing, with consequent effects on unemployment and the need for legitimation. The past is a learning experience, preparing for a new future but not one where the same solutions will work.

THE CULTURAL SUPERSTRUCTURE

A capitalist society is built on its economic base, in particular on the relations of production – the class system – within which all activity takes place. And a state is a necessary institution to capitalism. But there is much more to such a society based on materialism than its base and its state. Social life must be organized, and a great range of institutions make up what is known as the superstructure. Central to this is the family unit, within which economic and social life are sustained on a day-to-day basis, and within which the reproduction of society – physically and mentally – takes place (as in the inter-generational inheritance of property). Thus whereas the economic base sets the context for social organization, it is superstructural institutions such as the family and organised religions that socialise individuals into accepting their roles within society and legitimate certain divisions of labour – notably those related to age and gender, and in some cases race. As with the state, some such institutions are necessary in order to prepare people for their role in society. They identify what is normal and what is deviant – sometimes backed by the state (as in the establishment of certain religions and the banning of others); they 'suggest' that certain forms of consumption are desirable; they discipline people into the necessary constraints of working for capitalism – time-keeping, for example; and they provide people with identities, other than as workers. All that they do is acceptable to capitalism, indeed they are a major ideological bulwark for it. At some stages, they may threaten the path of accumulation, in which case the superstructure itself must be restructured, as, for example, with changes in the customs of inheritance where they produce too great a fragmentation of the ownership of land.

The family is the basic unit of social organization, therefore, and the state is the political unit. Intermediate to them, and playing a wide range of roles in the structuring of social life and reproduction, is a whole range of other institutions, some of which are separate from, but directly linked to, the economic organization but others largely independent of it. For some of these institutions, membership is assigned, and all individuals are allocated accordingly. With others, membership is by choice and people can (a) select whether they want to join certain institutions, and (b) select which others they want to be members with them. Such groupings (age, gender, race in the first type, for example; religion and a host of voluntary organizations, including political parties, in the second) participate in the structuring and restructuring of social relationships, thereby contributing to the reproduction of society (both base and superstructure) in a particular form.

Capitalist societies are extremely complex, continually changing entities, therefore. They are built on an economic base in which the means of production – land and labour – are unequally distributed. These means of production are put to work through the establishment of relations of production – a class system – which are a constant source of tension. The purpose of the work is to produce saleable commodities for profit. This too is a constant source of tension, since achievement of profits means cutting labour costs and overproduction (more is made than can be sold). Thus 'progress' is not linear, but cyclical. Boom periods – high levels of production, employment and profits – are followed by slumps – unemployment, poverty etc. – and restructuring then takes place to create new booms. There is, then, no long-term certainty of success, though all hope that it can be found.

Such a tension-ridden system is unstable. To put some stability into it requires a range of institutions that support and sustain it, and assist in its continuous restructuring, while being ostensibly outside it. Most important among these are the state and the institutions of social organization built around the family. The latter are the foci of daily life and reproduction; the former is the focus of the search for long-term stability though continual, relatively painless, change.

AND GEOGRAPHY

All of the discussion so far in this chapter has been about a model of a type of society with a particular economic base: capitalism. Nothing has been said at all about geography, neither the academic discipline itself nor the subject matter associated with it. The latter is usually taken as the areal differentiation of environments, the relationships between people and their environments, and the spatially organized inter-relationships among people. That this has not been mentioned so far could imply either that such areal differentiation does not occur, which is patently not the case, or that it is secondary to the operation, and therefore the understanding, of capitalist society. Again, that is not so. Geography is fully implicated in the operation of a capitalist society, as the discussion in the present section will make clear. Thus we must not assume that geographers simply map the outcomes of the operation of economic, social and political processes, so that their concern is with the epiphenomenal or surficial appearances of society only. Geography is part of society, indeed it is an integral part. The creation of a capitalist society involves the creation of a geography and that in turn is involved in the reproduction and restructuring of the society; society and its geography are in a continuous dialectic.

The Geography of the Economic Base

To illustrate this contention, I will begin with the economic base. The establishment of a capitalist organization must, of necessity, occur in a place or series of places. The materials used in the production process are place-bound, and must be either transformed there or transported to another place, where the other means of production, labour, will be applied to them. The output must then be sold, in a market – which is a place – and then consumed: much consumption is not immediate and complete (a building will be used for decades if not centuries; its furnishings for years if not decades; and so on) but is continuous – in a place. There are, then, three stages to the whole capitalist process: production, exchange and consumption. All of them occur in places, and for any one product both of the first two stages may occur in several places; consumption is usually, though not necessarily, in one place only. To the extent that the three are spatially separated, so different types of place emerge. Farms, for example, are dominantly places of production, as are factories. Many towns, or at least their centres and their *raisons d'être*, are dominantly places of exchange.

Other places, and certainly most parts of most towns, are predominantly places of consumption. And within each category there are many subcategories – different types of farm and factory, market place and suburb, and so on.

The nature of those places is far from unchanging. As noted already, capitalism is never static. Its dynamic involves a continual search for the ability to make more, more cheaply, and to sell more. That search is invariably spatial. To sell more, larger markets must be found, which means extending the area in which the products are sold – which in turn means extending the exchange procedures (for longer-distance trade), by requiring greater productivity in order to cover the costs. Thus space is a barrier to movement that has to be overcome as market areas are extended, either into virgin territory or into the market areas of other producers with whom competition then occurs. Similarly, the need to make the production process cheaper involves a continual search for new sources of raw materials which are cheaper than those already used, either because they are more easily won from the environment or because the labour involved is cheaper (in both cases with the savings in costs more than compensating for the greater costs of exchange).

A major element in the costs of production, and thus in the determination of profit levels, is the price of labour. Employers will frequently, if not always, be looking for ways of reducing this component. One way is to transfer production to places where labour costs are relatively cheap – particularly if those lower costs can be reasonably guaranteed for a given period (usually of at least several years) and are associated with a labour force possessing the requisite skills, readily accustomed to the disciplines of working practices, and unlikely to be in conflict with employers.

Every time there is a shift in the geography of production, therefore, the nature of at least two places changes: the place that is losing investment and employment and the place that is gaining them. As described above, the major shifts occur in the recessions of the so-called business cycle, when capital is moving in the search of new, profitable ventures. Such periods of restructuring are therefore major periods of geographical change. The search processes are inherently geographical, so that as the capitalist mode of production restructures itself so it restructures its places, by changing what is done where.

That restructuring will take place and that it will be geographical are both necessary elements of the capitalist mode of production. But neither the nature of the restructuring nor the nature of the geography is predictable, for two reasons. First, the restructuring is done by human agents, who are interpreting the need in the context of their understanding of the situation. Unless one can assume that all such agents have perfect understanding and are perfect decision-makers, you cannot predict what they will do. And since they are not, then prediction is impossible. (In any case, as Allan Pred, 1967, persuasively argued, perfect decision-making under capitalist competitive conditions is impossible, since it involves everybody outguessing everybody else.) Secondly, even if those assumptions were true, prediction would be impossible because it would have no baseline from which to operate. Every decision changes the context, and the world is never the same again. People change, because they are forever receiving new information, and assimilating it according to their own interpretations. And places change. Thus no decision takes place in exactly the same context as an earlier one; hence prediction is impossible.

Geography, then, is fully implicated in the dynamic of the capitalist mode of production, because that mode is place-bound in its operations and its continual restructuring processes involve the restructuring of places. Further, geography is implicated because what some see as its fundamental variable – distance – is a major constraint to economic inter-relationships. Distance is a barrier, the crossing of which takes time and consumes resources (thereby increasing costs). The expansion and restructuring of capitalism has been much concerned with reducing the height of that barrier – with annihilating space by time, to appropriate Marx's term – and the evolving geography of capitalism reflects how the barrier has been surmounted. Two aspects of it are particularly noteworthy.

First, the spatial expansion processes are particularly concerned with the search for cheap production (materials and labour) and for markets, that is, with production and exchange. Consumption is much more fixed in space, especially consumption based on profits, at the homes of the investors. The profits to be had from the production-sale nexus are filtered back to the investors, and those parts of the profits used for consumption purposes rather than for further rounds of investment are not consumed where they are produced. To the extent that there is spatial separation of production, exchange and consumption, therefore, so there is a real differentiation in the geography of the fruits of capitalism. And since capitalism is inherently spatial in its dynamic, so it is inherently spatial in the distribution of the fruits of that dynamic – profits. Some people, in some places, are major consumers of profits; others, elsewhere, are major producers. This creates what we know as the geography of development and underdevelopment. If development is indexed, as it usually is, by measures of consumption of goods and services, and that consumption is based on profits, then some places (in fact, some people in some places) can only develop by extracting profits from others; thus development in some places is necessarily counter-balanced by underdevelopment in others. (As stressed in chapter 2, the inequalities are relative, and the development:underdevelopment process can be associated, at least for a time, with increasing absolute levels of living in both places.)

The development:underdevelopment dialectic is to some extent the class structure of capitalism writ large on a spatial scale. But it should not be interpreted as simply a large-scale spatial phenomenon only. There are development:underdevelopment dialectics occurring in all places and at all scales, within the cities and countries at the development pole (often termed the cores of the world-economy) as well as within the places at the underdevelopment pole (the periphery). And thus there are inter-class conflicts at all spatial scales too, as people in some places seek to improve their position relative both to others in the same place and to others in other places. The dynamic of development is a spatial dynamic, taking place on a variety of local stages in a world-economy arena.

Secondly, distance is important because it can be manipulated in a variety of ways to promote the interests of certain groups, by the creation of spatial monopolies. For any investor, the guarantee of either or both of a supply of the needed resources and sales to a particular level will provide protection from the potential consequences of competition. Distance itself can provide this; if all other producers are too far away, then the costs of transporting their goods to a market make them uncompetitive there, and the local producers enjoy monopolistic or oligopolistic advantages. But to the extent that space can be annihilated by time, so those advantages are under threat. How, then, can spatial monopolies be sustained? One way is to make the annihilation processes difficult, if not impossible, as with the railway-builders of the nineteenth century who bought pieces of land to prevent competitors gaining access to a certain market. Another is to use political power to insulate local producers from outside competition, as detailed below in the discussion of the state.

It is not only in the spheres of production and exchange that space/distance is a resource to be manipulated for individual and group gain. It occurs in the sphere of consumption too, where inter- and intra-class conflicts are rife. Capitalist societies are characterized by private ownership of property, and part of consumption involves the acquisition of property, especially land and buildings. Indeed, some argue that such conspicuous consumption is necessary to the dynamic of capitalism, since it provides an ever-expanding market for the output of capitalist production. That consumption need not be of homes, their contents, and the means of getting to them, but there is no doubt – as Harvey (1975) has argued – that the drive for bigger, better-equipped homes has been crucial to the profitability of many capitalist societies in recent decades.

This form of consumption, in goods with long lives, is a form of investment by the individuals concerned, because they pay a price which covers the life of the product: they buy the home, rather than rent it. (Ownership is promoted by the state, too, as a legitimizing strategy; those who own property will be less likely to act to destroy their 'investment in the future'.) Thus they will want to protect their investment, to prevent external forces from generating some relative, if not absolute, decline in its value – not only because the quality of their lives will be affected but also, and increasingly importantly in the sort of mobile society which is also necessary to the capitalist dynamic, because it will mean that their ability to sell it will decline too. The built environment of residential areas, then, is the product of capitalist investment, and any threat to a part of it will be a source of conflict, between those who seek to benefit from the threatened change (often those involved in investment to restructure places in order to enhance profitability) and those who fear they will lose from it.

The built environment is not just an investment; it is the context for social reproduction, particularly family life. In a capitalist society, success can be guaranteed for very few; most must achieve it, by demonstrating their prowess. Such achievement is built on education, training, and experience; people must develop the right attitudes and values, must acquire the needed skills, and must obtain the entrées to the desirable positions. For these activities, the built environment is crucial because it provides the social context. Although the family is the basic unit of socialization it is insufficient; the skills and the contacts must be acquired in the wider contexts of school and neighbourhood. Thus families will want to be in neighbourhoods that contain the requisite contexts, and have the sorts of schools that provide the needed skills. In particular, they will want to exclude from their neighbourhoods those people who they think will harm the local environment.

The directly economic (investment in property) and the socialization requirements are linked, in that an area which is perceived as not good for socialization will not be attractive as one to invest in. Furthermore, an area whose quality is perceived to be declining will be one of disinvestment, and those with investment in it will seek to prevent that happening by halting neighbourhood change. Thus distancing is a major part of the creation of the built environment (what is widely known among geographers as the mosaic of urban social areas). The twin processes of congregation (like people grouping together in defined areas) and segregation (the exclusion of certain groups from defined areas) produce areal differentiation. This takes place particularly in class terms – though the overlapping of classes noted above means that there is some heterogeneity of areas – but not entirely so. Several of the other categorizations within the social organization of a society, notably that on racial grounds, may be a focus of distancing processes, as members of various groups within a society contest for the rewards and status that it has to distribute. Rarely is there explicit exclusion of groups from certain areas (South Africa is a clear example where that does take place, fully backed by the state), but manipulation of the market for property by a variety of means is undertaken to promote the equivalent of such exclusion.

The Superstructure

Geography, then, is fully implicated in the creation, extension and restructuring of capitalist societies, at all spatial scales and with regard to consumption (and the reproduction and restructuring of consumption) as well as to production and exchange. But capitalism was not created *in vacuo*, on an isotropic plain. It evolved out of pre-existing modes of production, and extended into and over (even removing) others. Thus the geography of those pre-capitalist modes has influenced the development of capitalism itself, as elements of the former, notably of their social and political superstructure, provided the contexts within which capitalism evolved.

Of all the modes of production that have existed, capitalism has been the most successful in annihilating space by time and integrating most of the earth's surface into a single world-economy. Some pre-capitalist societies were able for a time to control relatively large empires, combining economic and political power, though none of this was global in its extent and few were very long-lived. Most such societies were local in extent, small and, as a consequence, relatively stagnant. Some were entirely isolated from all others; most had some contacts with neighbours; and a few were linked through mercantile networks which supported the growth of large cities (notably in Asia) as centres of consumption. But even in such networks, the amount of contact was relatively small, and for the vast majority of the people concerned daily life was lived in a small, isolated spatial range or territory.

In each of these isolated societies, the local natural environment was the sole source of resources on which a livelihood could be based, and the mode of production that developed (in most cases based on either reciprocal relationships – that is, primitive communism – or, more likely, rank redistribution in which an elite was able to extract a surplus from the mass of the population through a concentration of psychological and military power) reflected an accommodation of the society to the local environment. This accommodation was far from determined, for the environment was both constraining and enabling; it provided the opportunities for, and yet limited, choices. People learned about their environments, developed strategies for coping with them, and established social and political systems linked to the seasonal and annual cycles of production and consumption. Societies chose how to live in their particular situations, and as those situations changed – with environmental deterioration, perhaps – so the societies accommodated the changes.

It was this mosaic of many local societies that provided the breeding ground out of which capitalism emerged. They varied enormously, in part because of the environmental differences but also because of the great variations in interpretations of how the environment could be used to provide a livelihood and how their societies should be organized. Some of these interpretations were much more likely to promote the transition to capitalism than others, it is argued, because their social and political systems generated the inequalities that provided the necessary capital for the first rounds of investment in labour power. (Coastal societies were better able to profit from trade, for example, and societies with primogeniture as the basis for inheritance were more likely to centralize wealth.) But it was individuals who created capitalism, by their selection of a mode of life.

Capitalism emerged in several places, slowly and contemporaneously and not independently. Because it was built on different pre-capitalist foundations, it developed in different socio-political contexts, so that the various capitalist societies retained many elements of those foundations. Out of these differences separate interpretations of the capitalist mode of production evolved. The economic imperatives of accumulation are the same in all, but the details of, for example, the class structure vary because they incorporate the pre-capitalist relicts (as in the large peasantry in some European countries). And the societies have evolved in different ways; the political party systems of West European countries differ substantially, reflecting separate developments of what are known as electoral cleavages around major areas of conflict. Thus capitalism as a mode of production is not the same everywhere. It has been interpreted in a variety of ways, to produce separate social formations, and these in turn provide separate contexts within which society has been reproduced and restructured.

As capitalism spread, and created the single world-economy, so the various social formations were imposed upon, or insinuated within, other pre-capitalist societies. This produced a great range of hybrids, of new social formations. And the contact between the capitalist social formations, as they competed for space and in places replaced each other (sometimes after war), created even more. This is marked in North America, for example, where Elazar (1966) has identified three 'cultures': the individualistic, based on free enterprise; the moralistic, based on community values; and the traditionalistic, based on elitist patriarchy. Each has its own spatial base on the east coast, and westward expansion has brought them into contact, with further hybridization as a result. This cultural geography was clearly implicated in the developing economic geography. The region of traditionalistic culture was not a centre of rapid accumulation and became relatively underdeveloped; it is now, during a period of rapid restructuring, the focus of much investment, balanced by disinvestment from the areas that boomed earlier.

The geography of capitalism, then, is a geography of social formations. The economic imperatives have been interpreted in a variety of ways, reflecting the pre-capitalist social organization either in the place or of the people(s) who colonized the place. These different interpretations mean that places differ in the ways in which capitalism is organized, is reproducing itself, and is being restructured. Capitalism is not just geographical in its operations, it is built on, and is maintaining, a complex and intricate geographical foundation.

The State is a Place

Lastly I turn to the state, already identified as a central and necessary institution for the capitalist mode of production. It differs from the other institutions, however, in that it necessarily operates within a clearly defined, spatially bounded territory. As Michael Mann (1984, p. 187) expresses it in an important essay: 'the state *is* merely and essentially an arena, a place' [his emphases], with the following characteristics: (1) a differentiated set of institutions (the state apparatus, see above, p. 35); (2) a centralizing focus of political relations; (3) a defined territorial reach; and (4) a monopoly within that territory of authoritative, binding rule-making powers and of the means of political violence. Why must the state possess the third of those characteristics? Why must the institution *be* a place, rather than just a place (an area on a map) being associated with the institution?

To answer this question it is necessary to look again at the role and functions of the state, and the power attached to them. In most pre-capitalist societies, the state possessed what Mann calls despotic power, the ability of the elite to exercise social, economic and political control without prior negotiation with those controlled; its only mode of legitimation was repression, linked to an ideology – frequently weakly developed and understood outside the main elite centres – which supported that role. As capitalism developed, so that form of legitimation became increasingly irrelevant and counter-productive, for two main sets of reasons. The first relates to the increased range of roles and functions undertaken by the state. Capitalism is a dynamic mode of production which is always changing as it reproduces itself. That change is dependent on changes in social relations and economic-political infrastructure, and so to promote accumulation the state has become more and more implicated in both daily life and the support of capital: it is a much bigger (and increasingly so) institution, with an expanding apparatus. The second set of reasons relates to the changing social relationships and the legitimation-consensus creation activities of the state. Restructuring requires acceptance if not consent, and the individuals concerned are likely to resent it, if for no other reason than that they are likely to suffer from it, temporarily at least, through unemployment. Furthermore, it is costly to them, because the state itself finds it necessary to invest in the restructuring. Thus the state needs to raise large sums of money from the people within its territory. It could do this through repressive policies, but the likelihood of success is not great over the long term because repression itself is costly, especially in the face of a well-educated population (whose education is provided, or promoted, by the state as a necessary part of its pro-accumulation policies). Legitimation therefore usually, though not necessarily, requires the building of consensus in such societies, and the exercise of infrastructural power which penetrates virtually all aspects of everyday life must be legitimized by consultation, normally through the processes that we know as liberal democracy.

Why does the exercise of infrastructural power necessarily require the state to be a territorial unit? States operate three forms of power, according to Mann – military, ideological and economic. With regard to the last, for example, it controls coinage and weights and measures, thereby providing a state guarantee of value for transactions, and it promotes rapid communication in the movement of messages, people and goods. With regard to military power, it maintains order within society and conducts relationships with other societies. And with regard to ideology it promotes consensus through the development of a state myth, a core set of beliefs providing rules to guide actions. Operation of these powers requires a defined territorial unit within which they apply. The military power needs a defined area to defend against outside aggressors, and within which all residents are bound by the laws that it enforces. The ideological power uses that defined area to identify the state with which people are associated; all states create a territorial identity, linking themselves and their citizens with their land. (Some ethologists claim that territoriality is a basic human need; others that it is a social creation.) And the economic power needs a defined area within which its rules hold sway – where its currency is the sole legal tender, its laws with regard to contracts hold; and so on.

The capitalist state, then, is necessarily a place, a defined area within which certain institutions predominate because such a place is prerequisite to the major roles of the state – to promote accumulation, to ensure legitimation of the capitalist mode of production, and to maintain social consensus and coherence. Other institutions in civil society may benefit from being linked with a place – a capitalist may benefit from a defined spatial monopoly, family life may benefit from a base in a defined area of land, and so on – but the spatial boundedness is not a pre-requisite to their activities. Indeed, for economic actors even the spatial monopoly is counter-productive in the long term, because the internal market protection it provides is countered by problems of infiltrating other markets, and therefore expanding. The capitalist system is now global, with no boundaries, and therefore to some extent the existence of territorially bound states presents barriers to its full articulation. For individual economic actors the state has some benefits, however, for it provides a territorially defined arena within which alliances can be built with other actors, to aid in the competition with others located elsewhere. Such economic alliances are frequently linked to the state ideology and are the basis for conflict between places which may erupt into physical violence; the economics of the state become the foundation of its geopolitics (Harvey, 1985).

The mosaic of state territories is not fixed, of course, because of the geopolitical role. Expanding economic alliances need an expanding state to support their extension into other areas. Areal expansion of the state almost invariably involves its military power, followed by its ideological. This was certainly the case in the imperial age, when the creation and maintenance of colonies was a major activity of the states linked to the most successful capitalist state formations. Today, military conquest is much less frequent; it is costly and difficult to justify. It has been replaced by economic and ideological conquest – neo-imperialism – which creates an environment for the operations of global capitalism without apparently interfering with the sovereign autonomy of the states.

The present mosaic of states developed out of the military and economic conflicts that accompanied the establishment of global capitalism. Not all of the states so defined were spatially coincident with the territories of pre-existing social formations. One of the goals of the new states was therefore to establish an ideological consensus, by a combination of repressive and socio-economic policies. Some have been more successful than others. For the latter, the problem of nationalism – the association of people with a common culture but without a defined separate territory – has led to conflicts within states, and the costs of handling these may well harm the ability of the states concerned to promote accumulation. The outcome of many of these conflicts is an alteration in the spatial configuration of the mosaic of states – most notably in the post-Second World War phase of decolonization.

The state exists because it is necessary to capitalism and it has particular roles as a consequence. How those roles are undertaken reflects on the individuals involved, however, because as in all other aspects of capitalist social formations there is no necessity for any particular action. The state is operated by human agents, who interpret their roles in their local contexts, using the tools of thought that they have developed there, and in turn creating the milieux for future operations. What is done is not predictable, therefore; why it is done is, however, since the state exists to do certain things.

As part of the superstructure of capitalist social formations, the state is subject to the economic base. Unlike other aspects of the superstructure, however, it has some autonomy from that base, because of its monopoly of certain types of power, particularly those associated with repression. Those in control of the state and its various apparatus can act to some extent independently of the economic interest of society, promoting policies that are favoured by the state elite (politicians and/or bureaucrats). The degree of autonomy is unknown theoretically, but it is certainly bounded; if autonomous policies are too antagonistic to the interests of capital, then eventually they will be undermined.

CONCLUSIONS

At the beginning of this chapter, I outlined a simple model containing three elements only: people, mechanisms and societies. The central position is occupied by people, who have created both of the other elements. The mechanisms that they create are the basic imperatives of those systems, their driving forces. The societies are the superstructures, the ways of organizing life in the context of those mechanisms.

Many mechanisms have been created in human history. Today one, the capitalist mode of production, dominates. One of its fundamental characteristics, as indicated here, is the necessity for continual change: stagnation means destruction. If the structure requires change, so the societies and the people involved must change. Thus the study of the capitalist mode of production is the study of change. For geographers this means study of the changing spatial organization of societies and the changing interpretations of the physical environment. This is not just the study of the outputs of the system, however, for geography is firmly implicated in the operation of the capitalist mode of production. Its dynamic is a geographical dynamic, and how that should be studied is the topic of the next chapters.

4

Story-Telling:
or Doing Geography

A proper story is like a river; sometimes it may be traced back to a
source in the hills, but what it becomes reflects the scenery through
which it flows. It has a history, and its history is marked by the
appearance of new incidents or new characters.

M. Oakeshott, 1983, p. 165

The previous chapter has outlined, all too briefly, the model of capitalist
society which informs both my approach to the practice of human
geography and the ideology within which I act as an individual citizen. With
regard to the latter, this clearly influences my views on the application of
human geography, the topic of chapter 6. Here, my concern is with the
practice of human geography.

The framework for the following discussion of my own approach to
human geographical practice is based on two foundations. The first is a
definition of human geography. This, of course, is a topic of some
considerable debate – leaving aside the views of those, like Michael Eliot
Hurst (1980, 1985), who believe that the discipline should not exist. And
yet, there is often general agreement that Richard Hartshorne provided a
viable definition in his *Perspective on the Nature of Geography* (1959,
p. 21) that: 'geography is concerned to provide accurate, orderly, and
rational description and interpretation of the variable character of the earth
surface.' This, of course, begs many questions because of its use of very
broad terms, such as 'accurate, orderly, and rational'. But these are
epistemological issues relevant to debates over the practice of human
geography rather than the purpose. For the latter, Hartshorne's emphasis on
'the variable character of the earth surface' provides me with an adequate
focus. It stresses *areal differentiation*, that places are different, and it is those
differences that I want to account for. Thus my attention is concentrated on
empirical questions, but this does not mean that my search for answers
remains at the empirical level alone.

The second foundation is concerned with philosophy and methodology, and follows on from the previous sentence. Three statements underpin that foundation.

1   Study of empirical 'facts' alone (I avoid the definition of facts for the moment) can provide descriptions of the world but can tell us little of how that world was produced. (The study of ripples on a pond which focuses only on the ripples cannot account for their existence. Nor can the study of the shape of the pieces of a jigsaw puzzle tell us much about the machine that cut them – D. Gregory, 1978, p. 99.)
2   Without an appreciation of the people involved in the production of the empirical world, how they are acting and why, it is impossible to understand the outcomes of their actions.
3   Without an appreciation of the mechanisms and societies within which people operate, it is impossible to understand the contexts in which they act.

These three clearly link back to the societies-people-mechanisms trilogy introduced above (p. 24), and it is important to keep them in the forefront of the discussion.

The three elements of the trilogy relate to the three levels in a realist philosophy of science, which will be discussed in greater detail later on. For the present purpose, it is necessary just to identify those levels and show how they relate to the practice of human geography. The three levels are: the *mechanisms* – the driving forces within the structures (known often as the real level); the *decisions* – the actions of the individuals (the actual level); and the *outcomes* – the results of those actions as empirically observed (hence they form the empirical level). These are linked, with the actual level forming the nexus (figure 4.1). The individual decision makers are not affected directly either by the mechanisms or by the outcomes, but rather by *interpretations* of one or both of them. Both mechanisms and outcomes structure the decision-making environment, but they do not determine the decisions – the interpretations do. And so our goal is to understand the interpretations, in that way gaining an understanding of how the outcomes have been created, and how they in turn influence their own reproduction and restructuring as influences on further behaviour.

The empirical world, in its many-faceted complexity, is an interpreted world, therefore; action is predicated upon meanings, not facts (as long as facts and meanings are not equated). This is so with any activity focused on the empirical world, including that of (natural and social) scientists, not least human geographers. As the previous two chapters have shown, what

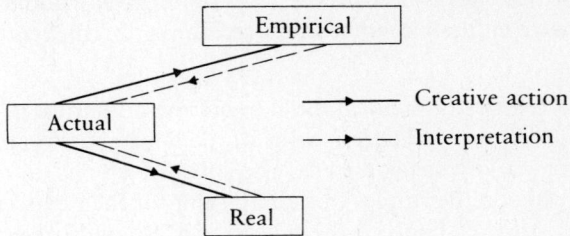

Figure 4.1   The three worlds of a realist approach to social science, and the links between them

we choose to study and how we choose to study it are properties of our interpretations, of what the empirical world means to us and how we believe the underlying structures operate. This creates particular problems for social scientists. The natural scientists' individual and group beliefs predispose them to study particular phenomena in certain ways. So do the social scientists', with one additional factor; social scientists are part of the empirical world that they are interpreting whereas natural scientists can separate themselves from it. In creating a discipline of human geography, therefore, we are interpreting *our* view of outcomes and structures, and how they are linked, to a wider audience which includes ourselves.

HUMAN GEOGRAPHERS AND PHILOSOPHY: A BRIEF EXCURSION

Until relatively recently, human geographers have not been beset by doubts relating to the philosophical orientation of their activities. They have, somewhat naively, identified their role as portrayal of the empirical world, as defined by Hartshorne, and have set out to do just that. In the last two decades, however, philosophical issues have loomed much larger and human geographers have questioned both the epistemology of the discipline (its response to questions such as 'What can we know?' and 'How can we know it?') and the related ontological issues (responses to questions such as 'What exists?'). Only when they have satisfied themselves on these issues, they concluded (or, at least, some did; most preferred not to bother), could they attack the methodological questions ('How do we obtain knowledge?'). A detailed consideration of how those philosophical issues have been tackled and debated is not needed here (it has been provided in the two predecessor books: Johnston, 1983a, 1983b). A brief outline is all that is provided, to lead into my own answers.

Varieties of Approaches to Human Geography

Just how many separate approaches to human geography have been considered in recent years is difficult to assess, for some of the literature at least is laden with unhelpful pedantic semantics (though I was once delighted to be associated with 'flying circus structuralism' – Eliot Hurst, 1980, p. 9 – since I tend to associate my efforts with the Goon Show rather than Monty Python; Dave Mercer's (1978) use of the term 'philomarxist' was kinder, but probably meant the same thing). Here, as earlier, I prefer to categorize the positions adopted into three main approaches, though with a slight change of terminology.

The *empiricist approaches* are, in general, the most naive since they are based on the assumption that there exists a pre-interpreted empirical world that can be observed, recorded and described by a neutral outside observer. This naivety was almost justifiable in what we like to term, pejoratively, 'capes and bays geography' which comprised little more than the statement of what is where, though of course the choice of what to catalogue reflected interpretations of what was worth cataloguing. It was certainly not justifiable in the successor, 'colonial geography', in which the goal was to describe the world in terms of phenomena of interest to those who wished to exploit it.

Empiricist work largely comprises unconsidered descriptions, biased in the selection of what and how to describe but with that bias unrecognized. It is built on what Giddens (1974, p. 2; after Kolakowski, 1972) calls the 'rule of phenomenalism', that experience of empirical events is the only basis for knowledge, reality is that which we perceive ('facts'). As Lacey (1976, p. 157) expresses it: 'We can only know appearances, but need no postulated unknowable objects lurking behind them, because belief or talk about such objects is really only a disguised form of belief or talk about the appearances themselves.' The empirical world is reality, and it contains within itself the key to its origins; we can explain appearances from appearances.

A particular form of empiricism is positivism, which carries the 'rule of phenomenalism' forward by arguing that understanding of empirical appearances can be gained by presenting them as particular examples of general laws. An early example of this within geography was the, rapidly discredited, approach to the subject known as environmental determinism, according to which the features of human geography could be ascribed to causes in the physical environment ('Basalt is conducive to piety' is an example etched on my mind). Thus the existence of one aspect of reality could be related to another (assuming, of course, that both basalt and piety could be unequivocally defined). This particular approach was soon ridiculed in its naive form, though researchers like Griffith Taylor continually emphasized its validity in terms of physical constraints rather than controls. But it remained, in a diluted form, in the practice of regional geography. This approach, later termed exceptionalism by Schaefer (1953), was based on a view of the world as a mosaic of phenomena-complexes, each of which was unique because it had a particular assemblage of phenomena. It was empiricist because it focused on appearances (although there were some attempts to identify an organic character to a region, based on a woolly concept of regional 'consciousness' or 'identity'); it was quasi-positivist, because implicit in many regional geographies was the primacy of the characteristics of the physical environment in determining the regional character – hence the debate over the relative merits of 'natural' and 'political' regions (Johnston, 1984a).

As a reaction to the naivety of regional geography, which many found increasingly unsatisfactory in the context of developments in other disciplines, human geographers returned to positivism – although not recognizing what they were doing until later – in their droves. This time they presented distance as the influence on (or determinant of) human action (see below, p. 132).

Within this form of positivism, which became known as locational analysis (after Haggett's classic, 1965, exposition), and to some as spatial science, two separate approaches were adopted. The first was firmly based in the hypothetico-deductive set of procedures generally associated with natural science. Certain attributes of behaviour were assumed (such as cost-minimization and profit-maximization) and these were used as the basis for the derivation of normative models of the spatial organization of society, which were then tested empirically. The other was more inductive, seeking to identify regularities in spatial behaviour without assuming that it obeyed particular rules; the rules were to be discovered from the empirical analyses, but they were to be general rules, not those applied only by the people studied.

The second set of approaches is generally referred to as *humanistic*. These deny the existence of an empirical world outside the observer, and emphasize the role of meanings in the creation of the environments in which people live. Thus it is not possible for a human geographer to be a neutral observer, because what one observes is a consequence of the meanings that one applies. There are no neutral 'facts'. Nor is it possible to understand meanings through observation alone, since meanings are mental not physical constructs. And, thirdly, one cannot assume that laws of behaviour can be developed, since this assumes that meanings are both shared and fixed.

The central feature of humanistic approaches to human geography is subjectivity – not the subjectivity of the research itself (despite some claims to the contrary) but the subjectivity of meanings. The world in which we act is not a world of meaningless things (what Kirk, 1963, calls a phenomenal environment) but a created world of meanings and interpretations (the behavioural environment, according to Kirk). The two overlap, of course, but must not be equated. Thus to understand how people act, one must appreciate the context in which they act – the behavioural environment, which is a repository of the meanings that they attach to phenomena; one must study their subjectivity objectively.

Although the general thrust of the humanistic approaches was acceptable to many, the methodology to be used in its application was uncertain, in part because of insufficient exploration of the epistemologies of various forms of humanistic research. Several epistemologies were advocated. One such was idealism, associated particularly with the writings of Leonard Guelke (e.g. 1982). According to Guelke's presentation, human action is predicated on personal theories, so that the task of the human geographer is to identify those theories. Thus the human geographer approaches the subject to be studied without any prior theory; all that is needed is an appreciation of the theories of the subject(s). With this, 'the explanation of an action is complete when the agent's goal and theoretical understanding of his situation have been discovered. . . One must discover what he believed, not why he believed it.' (Guelke, 1974, p. 197) With reference to figure 4.1, therefore, the human geographer who adopts the approach of idealism is concerned with the interpretation links between the actual, the empirical and the real, *but not* with how they came about.

Of other humanistic approaches the one most frequently canvassed among human geographers is phenomenology, about the nature of which there is some confusion (see Pickles, 1985), because of different presentations and interpretations (i.e. phenomenology itself exists only as individual interpretations!). In its basics, phenomenology shares with idealism a focus on the life-world, the behavioural environment that contains the meanings which predispose action. It differs on two points, however. The first relates to methodology and reflects the separate origins of the two philosophies. Idealism is particularly associated with historical scholarship (notably the work of R. G. Collingwood, on whom Guelke relies heavily), and thus can only gain access to the theories behind action indirectly, through the relicts of the period/events being considered. In this sense, the practice of idealism is what it studies; the building up, by the historical scholar, of personal theories of the subject matter. Phenomenology, on the other hand, is more closely linked to the contemporary concerns of sociology and psychoanalysis, and so seeks *verstehen* (empathetic understanding) though procedures (such as Freud's method of cooperative encounter and exploration) involving inter-personal contact. The second difference relates to attitudes towards the existence of the personal theories and meanings that are identified. Idealists, according to Guelke, are not concerned to know why people believe what they do. Some phenomenologists are, however, because they believe that exploration of meanings will reveal essences – those aspects of human consciousness that control the allocation of meanings.

The third group of approaches is popularly known as *structuralist*, thereby implying a concern with structures and mechanisms. The essential feature of all structuralist approaches is the belief that meanings and actions are bound together in logical structures, the existence of which cannot be apprehended empirically but can only be appreciated intellectually – that is, by the construction of theories. Thus all behaviour is the consequence of underlying, unseen, mechanisms that reside in structures; one cannot observe its generation, only account for it by a logical theory.

Structuralist approaches vary in their presentation of how behaviour is generated. Some are quasi-determinist in their presentation, such as the structural marxism in which, according to Duncan and Ley (1982, p. 32), 'The whole determines the nature of the parts', with the wholes as 'reified entities such as capital and the mode of production' (p. 38) and the individual humans as mere puppets: 'the means of implementing the goals of the formal cause' (p. 38). Such structuralism, they claim, is teleological. But other approaches within this set can more readily be termed possibilist, since the mechanisms are presented as both constraining and enabling. Piaget's developmental psychology, for example, presents child development as the interaction between an organized intellectual system – the human ability to assimilate material into mental schema and to modify those schema as a consequence of that information – and an external world, as in learning how to find routes through a maze: what is not clear is the degree to which the end is given, and full development of the interaction will always produce the same result. Linguistics, as presented by Chomsky, and anthropology, as argued by Lévi-Strauss, are clearly possibilist. They identify basic mechanisms within the human intellectual system (rules of grammar in the first case, and of social interaction in the second) but accept that those rules will be interpreted in different ways (producing different languages and incest taboos, for example), in exactly the same way as cricket captains interpret the rules with regard to the location of fielders (although those rules can be changed by dictat from the great structuralists at Lords – the MCC).

The goal of a structuralist analysis is to identify the mechanisms, by the construction of coherent theories that are consistent with the outcomes. In some analyses, those mechanisms may be part of the human intellectual system. In others, such as the structural marxism criticized by Duncan and Ley, they are human creations, not aspects of the human (which may mean that an even deeper structure should be sought – those aspects of the human intellectual system that have produced the capitalist mode of production, for example). But in either case the focus is on the mechanisms, not the empirical outcomes. In terms of figure 4.1, all the links are studied, because it is important to know what the mechanisms are, how they were put in place, and how they influence interpretations.

Putting Them Together?

Many debates begin by being polarized and end with the search for middle ground, based on the 'best' elements of the various initial premises. This has been so with the philosophical issues that have been debated in human geography in recent years. The poles are those of determinism (whether of the positivist or the structuralist form) and voluntarism – pre-ordination or free will, teleology or uncertainty (see Gregory, 1981, and Thrift, 1983). The middle ground seeks to combine elements of both, to bring the various philosophies together in some hybrid form.

No discussion of those proposed hybrids is provided here (see Johnston, 1983b), for most are as unsatisfactory as the polar positions on which they draw. This is because they are based on philosophies that are themselves unsatisfactory, since they do not accord with the three statements set out at the beginning of this chapter. All empiricism is rejected because it postulates separation of an 'empirical reality' from the minds of both the observers (i.e. the researchers) and the actors within that reality; meanings either are ignored or are treated as observable, equivalent to inanimate things. Positivism is further rejected because of its goal of identifying laws of human behaviour, which implies a freezing of a dialectic and unpredictable process of change and denies freedom of action to the human actor: it is, ultimately, a determinist philosophy based on an untenable attitude to a subjectively constituted reality. Idealism is rejected because it promotes a separation of agent from structure and, as is made clearer in existentialist philosophy, is built on the assumption that people make themselves; individuals are torn out of context, and the nature and influence of that context are disregarded. Some structuralism is more acceptable, because of its foundation in the study of the underlying mechanisms. But, as Duncan and Ley's (overstated) critique makes clear, some uses of structuralism are unsatisfactory because they are, like positivism, ultimately determinist, giving the human agent no freedom of action.

THE REALIST ALTERNATIVE

Although the three sets of approaches that have attracted human geographers have been rejected above, nevertheless some aspects of each are relevant and needed. Structuralism without any deterministic overtones is certainly required since, as stressed many times so far in this book, there are mechanisms within the capitalist mode of production that are not apprehendable but which exist, because we have created and sustained them, and which provide the ultimate controls over the way we live. And humanistic approaches, without their extreme voluntarist elements, are similarly necessary, for they focus on the crucial role of meanings as influences on action. Empiricism, and positivism, have less to offer in this general sense (although, as the next chapter shows, the critique of positivism has frequently been overstated and misdirected).

The search for a viable approach to human geography must, therefore, focus on some combination of what are identified as the valuable elements in structuralist and humanistic thinking, *in order to advance understanding of the empirical world*. Our focus as human geographers is, I contend, on that understanding, and our viewpoint accepts the anti-empiricist contention that there is no single empirical world but many overlapping yet discordant worlds of meanings. We seek to interpret those worlds, to appreciate what they are, how they have been produced, how they are being reproduced and/or restructured, and how they influence action.

The approach that I advocate is the one outlined in figure 4.1, which is linked to the societies-people-mechanisms trilogy of chapter 3. Its first level is the empirical, or the world of experience. But experience is an interpretation, and interpretations can only take place within a framework of theory. As Piaget's structuralism makes clear, individuals are provided with the tools for thinking, with an intellectual system, but not with the materials. I may observe one person killing another: this observation has meaning for me because of the moral code that I have learned and which I apply in such circumstances, a moral code which may differ from that applied by others and which I express using particular terms in my language – such as murder – which have defined meanings, that I share with others. Thus in order to experience, and to act in, the empirical world I need both intellect and concepts: the former is part of my genetic structure and is given to me; the latter I acquire and organize from my context.

The empirical world that I experience is the consequence of actions, past or present. Those actions are undertaken by people with intellects and concepts, just like me; they may differ somewhat from me in their intellects (their natural ability to handle information) but, much more importantly, they possibly differ from me in the concepts that they use. Thus I may interpret the killing that I observe as murder, with all that the term implies in my moral code: the individual perpetrating the killing may perhaps see it as self-defence, or as a necessary propitiation to an angry god. The interpretations are out of phase, because the creator of the event gives it a different meaning from that of the observer. They are living in different conceptual worlds. Neither is right, or wrong. My interpretation of it as murder is both a statement of the meaning that I allocate to the event and a further piece of information which is added to the store of knowledge on which my behaviour is based. The killer interprets it differently.

For the researcher, both interpretations are relevant to an overall study of the event: without understanding the thoughts behind the killing, it is impossible to appreciate why that event took place; without understanding my attitudes to it, it is impossible to appreciate any reaction that I might display. The researcher must understand both people involved. (This provides a further illustration of the poverty of empiricism: if I were to study that killing in the positivist sense, perhaps as part of a work on the geography of crime, I would present it as one occurrence of the category murder, therefore imposing my interpretation. Knowledge of it as my interpretation, and no more, is valid, but if I present it as the sole interpretation, then I am distorting the action.) Hence the level of the actual is separate from that of the empirical; the latter is concerned with the world of experiences, and the former with the world of events.

Experiences and events do not provide an enclosed, mutually reinforcing system, however. Events must be the actualization of something, putting an idea or a motivation into action. It is the structures that are actualized, the mechanisms that underpin the operation of society. Again, this involves interpretations. The killer who is making a sacrifice to the gods does so because this is how she or he interprets the religion that guides action, that religion being part of the superstructure of his or her society. My interpretation of it as murder reflects the moral and legal code of my society, which itself will have a religious base, suggesting that the two superstructures do not agree. But beneath the superstructural level is the economic base, that requires a superstructure within which the mode of production can be actualized. The creation of that superstructure is itself the result of many actualizations, continually reproducing the institutional framework within which events take place. The events need not be directly related to the mechanisms of the economic base, for not all actions by any means are even indirectly related to the imperatives of the mode of production. (Racial discrimination is not necessary in a capitalist society, for example.) Many are involved with activity within, and maintenance of, the superstructure, but it must be recalled that the superstructure exists because it is necessary to the base. Thus different superstructures may have different religions with separate value systems, hence the different interpretations of the killing. To appreciate fully why those different interpretations exist, we need to know not only how the two religions have differed, but why a religious superstructure is needed within the societies concerned. (Of course, with geographical restructuring social formations are often mixed together, producing new ones with disparate parts.)

Crucial to all levels of the realist schema is the individual actor, or human agent, because in human geography the actions studied are those of knowing agents: the empirical world is created and interpreted by knowing agents; the base and superstructure are created and are interpreted by knowing agents. This gives us the model of figure 4.2, which contains six

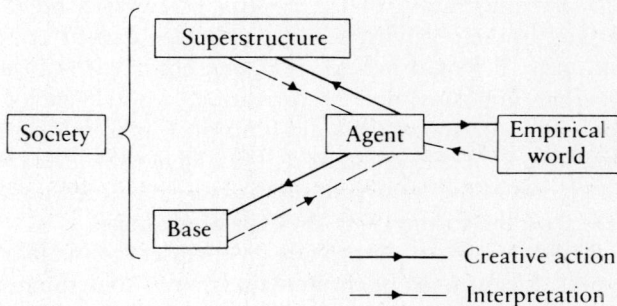

Figure 4.2   Base and superstructure in the realist schema

sets of interactions, three of them interpretations and three creative actions. All are operating simultaneously and continually, separately for every actor. This could suggest both anarchy and the basis for a voluntarist philosophy: neither is correct, because those interactions take place in context, and it is the context that binds people together.

The idea behind figure 4.2 is a simple one, and yet it is crucial. As Derek Gregory (1981) has shown, many of the implicit models of human action that have underpinned human geography in recent decades (what some call 'models of man': van der Laan and Piersma, 1982) are based on an incomplete set of interactions. Thus in reification models, society comes first and creates the individual, with no reciprocal impact: society (base and superstructure) is *sui generis*. (Gregory cites Durkheim and some neo-marxists as typical of this model.) Voluntarist models, as in the works of Max Weber, lack the society-individual links: society is no more than the sum of its member parts. Dialectical reproduction, associated with Peter Berger, has society-individual-society interactions, but only after society is established: society somehow is antecedent to the individual. Only structuration, an approach espoused by Jurgen Habermas and Anthony Giddens, begins and ends with the individual: individuals create societies, which influence individuals, who recreate societies etc.

Structuration is an approach which has received considerable attention from social theorists in recent years, largely as a result of Giddens's advocacy, and it has been promoted as a viable approach for the study of human geography: indeed, Giddens (1983, 1984) has incorporated elements of human geography in his most recent presentations (though see Gregson, 1986). The essence of structuration is that of figure 4.2. People live in structures, organized sets of rules and resources that they have created and which, as they enact them, they reproduce. Thus agents and structures are not independent (there is no dualism of structure and agency, in Giddens' terms). They are interdependent, forming a duality: 'the structural properties of social systems are both medium and outcome of the practices they recursively organize' (Giddens, 1984, p. 25). We live through the rules of the society of which we are part, but in living by those rules we are part of the rules; they are reproduced, not necessarily in exactly the same form, as we enact them. (Language is an excellent example of this. It is a set of rules for communication which we use. As we use them, so we reproduce them, hence it is a 'living language' constantly reproducing itself through the medium of the users – and hence, also, it is different from a biological organism which is reproduced without any action on the part of a knowing agent. As we use the language, too, we change it in subtle ways, to accommodate changes in the context to which it is applied: new words are invented, old ones are given different meanings, new sentence forms become 'acceptable', and so on.)

Although the concept of structuration is readily accepted, along with other models of the type presented in figure 4.2 but which do not necessarily appropriate the term, its application is less obvious (as Gregson, 1986, argues). How does one undertake empirical work in the context of those three mutual interactions shown in figure 4.2? Is it not necessary to break into that recursive set of relationships, to start somewhere and therefore give either structure or agency, society or individual, primacy?

DOING RESEARCH

The argument that I will advance here is that structuration provides a framework for research, not a methodology. It is the equivalent of an academic ideology, a core set of beliefs with which is associated a set of action principles (the terms are taken from Scarbrough, 1984; the orientation is similar to that of research programmes as outlined by Lakatos, 1978). Any one piece of research does not look at the system of figure 4.2 as a whole, but is related to some part of it. But the goal of the research is to contribute to the understanding of the system, building up an appreciation of the whole from a consideration of its parts. (To some, there is a difference between research, which is the identification of new knowledge, presumably empirical, and scholarship, which is the articulation of research findings into coherent bodies of understanding. To many this is an artificial distinction, not only because the articulation itself contributes new knowledge but also because it separates researchers from scholars when the two roles should co-reside in the one person. Nevertheless, it is possible to use material from one research programme to inform a synthesis based in another. Carter, 1982, and Yeates, 1983, for example, have criticized my attempts at creating a synthesis for urban geography and illustrating it for the USA (Johnston, 1980b, 1982) because they draw on research materials from 'another paradigm'. This is invalid criticism, however. All I have done is to take the 'text' – the term is developed below – from pieces of research, appreciated the context in which that was done, and used it, with no distortions, for my own text: empirical description is value-laden, as already stressed, but if the values are appreciated its output can be assimilated into other value systems.)

A researcher may, then, choose to focus on the real, the actual, or the empirical. With regard to the last two in particular, the problems that this may raise relate to the twin concepts of abstraction and chaotic conception. Abstraction, I have already suggested, is necessary for the conduct of research; a part must be separated from the whole. But it should be a rational abstraction, a part which can sensibly be studied in isolation and the conclusions of which study can readily be incorporated into the whole. If it is not, then it forms what is called a chaotic conception (Sayer, 1984, p. 127); it may be so heterogeneous a concept that valid conclusions cannot be drawn from it (Sayer uses the example of the term 'service industry' which covers so many disparate parts that they should be separately studied), or only a part of a homogeneous concept is abstracted, as in the simple definitions of 'working class' using a single variable only (usually occupation) in many voting studies (see Dunleavy and Husbands, 1985).

Avoidance of chaotic conceptions means that the research, whether at the level of the actual or the empirical, must be carefully structured within an overall theoretical framework. Concepts such as 'service industry' and 'working class' are widely and variously used. But for a piece of research which aims to contribute to knowledge they must be carefully defined, and the definitions must be relevant to the framework within which the results are to be set. Service industries include merchant banks and plumbers, which undoubtedly have different geographies. To lump them together in a single category and to describe its geography will reveal little about the spatial organization and restructuring of society; the contexts and actions of merchant bankers are very different from those of plumbers.

The goal, then, is to develop a theory of society out of which viable actual and empirical research can be undertaken, and which in turn feeds back into that theory, which must itself always be changing as the society it studies changes. Thus Scarbrough's (1984) analysis of *Political Ideology and Voting* in Britain groups electors according to their ideological orientations and explores the links between those and voting behaviour over a sequence of five elections. Although there is a general discussion of the concept of ideology, nothing is said about, for example, either the need for an ideology in a capitalist state such as Britain or why political parties promote – in every sense of that term – separate ideologies. As such, it could be dismissed as a piece of empiricism, with strong voluntarist overtones. But the conclusions are that:

> we would do well to pay close attention to the role of political ideology in determining the affairs of the community. And as it lies with electors to determine which version of affairs should prevail, we

should not hasten to dismiss political ideology from the minds of electors. They may not know as much about politics as party leaders ... But electors are members of the same political community as politicians; we suggest that they share many common understandings about their community. Expressed in political ideologies, it is these common understandings that imbue democratic politics with its life and its direction. (pp. 220–1)

Thus it is not a question of consumer sovereignty in the traditional sense of behavioural study, it is rather an issue of people learning common orientations and then acting on them. How those orientations are learned, and why they are promoted, is outside the orbit of the study, but what has been reported relates to a rational abstraction from the model in figure 4.2: the results can be incorporated to a wider programme of understanding the empirical world – the pattern of votes – and the actual events – the decisions on how to vote.

Although the view of society presented in chapter 3 and the approach to its understanding outlined in figure 4.2 is holistic, therefore, it is possible to make rational abstractions from the whole and undertake research – into the theory of the real, the behaviour involved in the events of the actual, and the empirical outcomes – without violating that holistic framework. Valuable research is based on rational abstractions. It also avoids the problems that are associated with positivism and voluntarism. With the former, it is easy to fall into what I have termed (Johnston, 1985b) the generalization trap, assuming (implicitly if not explicitly, often because of the language employed) that the findings of a case study are transferable in time and space (see Baker and Gregory, 1984). With the latter, there is a danger of falling in the other direction, into a singularity trap, of assuming that the findings of a case study have no relevance to any other area of knowledge. Events are unique, but they are not singular, because they are responses, in context, to the driving forces of society.

RESEARCH IN CONTEXT

For human geographers the critical point to be drawn from the previous section, which gives them a clear research niche in the current academic division of labour, is that events *'are responses, in context'*. Geographers, and others (such as Giddens, 1984), argue that one of the most important contexts is *place* (Giddens's term is 'locale') – hence the titles of arguments that *Geography Matters* (Massey and Allen, 1974) and 'Places matter' (Johnston, 1986a); Pred (1984a) expresses the same simple idea in a more convoluted form: 'Place as historically contingent process: structuration and the time-geography of becoming places'. (Others argue against this. Dunleavy (1979), for example, has attacked those, including geographers, who argue that there is a local contextual effect in voting behaviour, that, as Miller (1977, p. 65) expresses it: 'The effect of the social environment may be explained by contact models: those who speak together vote together.' Dunleavy's attack on this relatively naive spatial determinism has some force (see Johnston, 1985a), but his counter-position (expressed in Dunleavy and Husbands, 1985, p. 18ff.) that voting is conditioned by group – i.e. class – interests lacks force because he fails to discuss how those group interests are fostered. My argument is that they are fostered by socialization *in places*.)

How, then, do we structure research which takes account of this geographical case? Andrew Sayer's (1984, p. 129) schema showing the relationship between abstract and concrete research points the way (figure 4.3). At the abstract level one has the theorization of what he terms 'transhistorical necessities' (e.g. the need for people to reproduce themselves through the consumption of food), through to the specific structures put in place to meet those necessities (e.g. the capitalist mode of production). Those structures contain mechanisms which are operated concretely – that is, they are actualized and produce empirical worlds – in particular contexts, what he terms 'contingently related conditions', such as 'the particular kinds of technology available, the relative power of capital and labour and state intervention' (p. 130). Further, 'No theory of society could be expected to know the nature and form of these contingencies "in advance" purely on the basis of theoretical claims.' Thus the move from the abstract to the concrete involves shifting from transhistorical to historically specific claims; to understand how the structures have been transformed into empirical experiences one must appreciate the historical, *and geographical*, context of the transformation.

That the contexts are place- as well as time-specific is made clear by the arguments of time geography. These show that, because of a combination of capability, coupling and authority constraints, people's lives are packed in time–space prisms. Places matter because for most of us, for most of the time, we are confined to particular locales. It is within those places that we learn, that we assimilate and accommodate information; in the early years of our lives, the locales are usually very restricted and the whole process of socialization is very much constrained. It is, of course, from people that we

Foundations of historical materialism
(e.g. concepts of people and nature)

Transhistorical claims
(e.g. nature of human labour, social
relations of production)

Historically specific abstractions
of necessary/internal relations
(e.g. capital—wage-labour)

A
b
s
t
r
a
c
t

Tendencies/mechanisms operating in
virtue of necessary relations:
$x_1, x_2 \ldots x_k$ (e.g. law of value)

Contingently related
conditions*
(including other tendencies)

C
o
n
c
r
e
t
e

Synthesis of tendencies and conditions
('unity of diverse aspects') to form
concrete concepts: $z_1, z_2 \ldots z_k$

Conjunctures
(within 'open systems')

*The theorization of these, and their explanation by means of abstraction,
is often not the sole prerogative of marxism

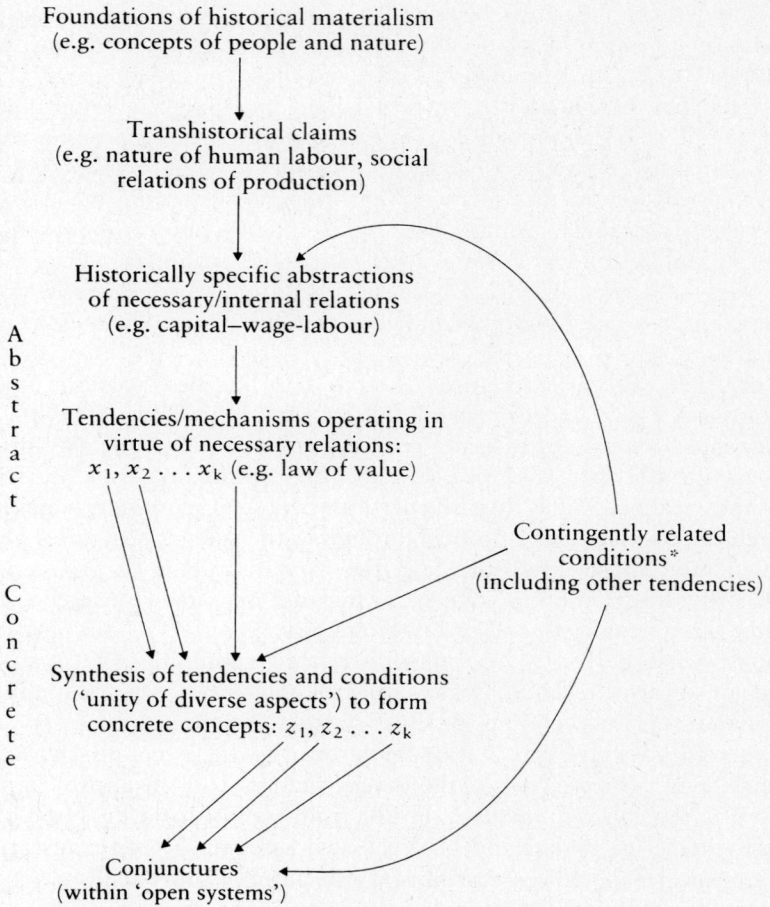

*Figure 4.3   The relation between abstract and concrete (reproduced, with
permission, from Sayer, 1984, p. 129)*

learn, that we obtain the concepts that are manipulated by our intellects, and with which we interpret the real and the empirical. To the extent that space has been annihilated by time in the sending of messages, so the locale's boundaries are widened; even so, we have to learn what those messages are and develop attitudes towards them – even whether we will bother to receive them – so people in places are crucial to how we learn, at least initially.

Places are the contexts in which we learn and live, so the contents of our particular places (the ruling ideas etc.) are crucial influences on our behaviour, on the worlds that we experience and make for ourselves. But they are not autonomous; they are necessary abstractions. As Peter Taylor (1981a) expresses it, the real place is in fact placeless, for the reality within which we live is the global world-economy. For most of us, however, the empirical place, what he calls the 'scale of experience', is the locale within which we live – for most of us a part, if not the whole, of an urban area. These places define the quality of life that we experience, and are where the problems of the scale of reality – the global world-economy – are apparent to us. In between the two is what he calls the 'scale of ideology' which is the territory of the state. We are socialized in a place that is smaller than a state but, as a consequence of what Lipset and Rokkan (1967; see also Urwin and Rokkan, 1983) called the 'national revolution', we are also socialized into the belief that 'our place' is merely a part of a larger place, the nation state, to which we owe loyalty. It is at the level of the nation state, for example, that most political ideologies are formulated as responses to the operations of the global economy and to which we learn to react through the ongoing processes of political socialization at the scale of experience. In this way, Taylor (1985a) produces a *Political Geography: World-Economy, Nation-State and Locality* (see also Short, 1984) which links the spatial expression of the real world, and its globally oriented economic actors, to the local, social world via the politically ideological world, the state.

Whatever the scale, action reflects the conjoint interpretion of base and superstructure (figure 4.2), with the interpretation being partial because it is contextually based. People in different places are probably operating in different contexts, since the majority of their information sources, both the relevant cues for a particular behaviour and the background knowledge with which they are assimilated, are local. But people in the same place may be operating in different contexts too. Within a town, for example, there are many distinct social networks. Most overlap with others, so that indirect contact between any two people could occur, but the majority of people live in a segmented world, defined by age, gender, race, religion, language, occupational position and status etc., so that theirs is a particular context within a more general one. Further, some people – usually though not exclusively defined by occupation and economic status – have much wider (spatially as well as numerically) social networks than others (as Pred, 1984b, makes clear in his description of nineteenth-century Boston) and are drawing on different contexts in their decision-making.

Two types of empirical work are characteristic of much contemporary human geography. The first is *the description of the world of experiences*, which is usually accompanied by an attempted account for what has been described. Given the spatial scales at which geographers work, especially in comparison with anthropologists and, to a lesser extent, sociologists, most of these descriptions refer to large aggregates of data – the outcomes of many events, as in descriptions of residential patterns, migrations, the distributions of factories and so on. Many of them rely on data collected by others, usually for purposes other than the human geographer's, and so the user is tied to the concepts of the collector and the interpretations of the empirical world employed (hence the choice of census questions, for example). Such data are limited in value, especially if they are based on chaotic conceptions, but their analysis and careful interpretation can provide insights to the reasons for their creation and information on the contexts within which future decisions will be taken. The generalization trap is, of course, ever-present, and there is always the danger that the analyst's own context will provide a false impression of how the actors who produced the data perceived their worlds.

The second, and potentially more valuable, type of research is *the description of particular actions*, in which the observer looks not just at the outcomes but at the reasons for the events producing the outcomes. Much of the work conducted in this mould suffers from one or both of the following: a positivist inclination, typical of much work in behavioural geography, which although based on data collected from individuals treats people as members of externally defined categories (age, gender, occupation etc.), so that it is the researchers' interpretations of the world that are dominant, not the actors'; and a reliance on data that can be collected by questionnaire, which is often only superficial in exploring the theories behind action and which again imposes the researchers' concepts on the respondents.

Both of these types of work are of value in providing material to be slotted into the holistic framework, given that, as indicated earlier, holistic studies are extremely difficult, if not impossible, to achieve. But their engagement with their subjects is relatively slight, and the degree to which they provide an understanding of how and why people act in the way that they do is also slim. And yet methodologies are available with which such insights can be obtained, many of them deriving from the ethnographic studies of the 1920s/1930s Chicago School of Sociologists. (As many have pointed out, e.g. Jackson, 1985, geographers have drawn heavily on the human ecology elements of the Chicago School's work, and have largely ignored the ethnographies.) They are time-consuming and, in many senses, more difficult than the more characteristic data collection and analysis methodologies, because they involve establishing empathetic inter-relationships with those being studied, preferably over long periods (see Jackson and Smith, 1984). But they provide access to the thoughts and theories of the individual, without which full appreciation of both their and our empirical worlds is impossible.

The importance of understanding individuals cannot be stressed suffi-ciently, for whatever the context it will only produce a particular outcome if one or more people act in a certain way. However favourable the contingently related conditions (figure 4.3), if they are not actualized then a particular outcome will not ensue (as Sayer's, 1984, gunpowder example makes clear). Conversely, however unfavourable those conditions, one action is sufficient to produce an 'unexpected' outcome. The environmental context provides the gun, but it needs an individual to point it and to pull the trigger. Manuel Castells (1983) makes this clear in his analysis of why San Francisco has developed as the centre for gay culture in the United States. Many of the local circumstances were favourable preconditions:

> [during] the Second World War, San Francisco was the main port of embarkation and disembarkation for the Pacific front. An estimated 1.6 million men and women passed through the city: young, alone, suddenly uprooted, living on the edge of death and suffering, and living with people of their own sex. The average ten per cent of homosexuals found in all human populations found themselves more easily and rapidly in this context. Many service men and women were discharged from the military for homosexuality . . . they stayed in the city . . . They met in bars . . . the focal points of social life for gay people; and networks were constructed around these bars; a specific form of culture and ideology began to emerge. (p. 141)

The conditions were propitious, but they probably were in other places too. It was the bar-owners, who were prepared to foster that gay culture and the creation of a place, who triggered the reaction from those conditions through, for example, encouraging 'the initiative, courage, and imagination of . . . José Sarria, a famous drag queen . . . [who] for 15 years . . . dressed as Madam Butterfly and . . . [delivered] a "sermon" about homosexual rights' (p. 141).

The bar culture produced an incipient place, therefore. And then in the late 1960s the development of the hippie counter-culture, also centred in San Francisco, provided the contingently related conditions for it to develop into a neighbourhood community. Following this, in 1973 one of its residents – Harvey Milk – decided to mobilize the community's political strength by running for supervisor; he won election in 1977, and in the backlash against that success, and of the general visibility of the 'deviant' gay culture, was murdered in City Hall in November 1978.

The gays in San Francisco created a place, therefore, both a local culture and a political milieu. The latter was the result of mobilizing the electorate around a particular cause – creating an electoral cleavage, in the normal jargon. Such community-based strategies are common in many political systems (Johnston, O'Neill and Taylor, 1985). A party may deliberately set out to mobilize the electorate in an area, or an individual may do so, as in many recent British by-elections. By doing that, they alter the nature of the place, and so change the context for future contests. (As Husbands, 1983, shows in his analysis of the National Front in Britain, the collective memory is often long-lived; once created, a place tends to survive.)

Detailed ethnographies are few in human geography. One of the small number completed illustrates their value. David Ley's (1974) *The Black Inner City as Frontier Outpost* contrasts the insiders' and the outsiders' views of the Monroe district of Philadelphia, showing how people structure a place in order to live in it. But that is not just an isolated piece of reportage. Through his text on *A Social Geography of the City* (1983) he is able to use such work to provide general insights into urban milieux; it is the proper use of a case study. Further, and more importantly, he is able to use it, in the context of constitutive phenomenology (Ley, 1977), to illustrate how we create and live in taken-for-granted worlds. Because the world of experience is complex and ever-changing, we need to simplify it in order to survive. To avoid detailed evaluations every time we have to make a decision (how to interact with a particular individual; how to get from A to B etc.) we create systems of ideal types, which structure our habitual behaviour; instead of having to take decisions, we make unconsidered choices. Those types represent the human, built, and physical environments that we live in, and when we encounter new elements we allocate them to the relevant types and act accordingly.

Here we have a clear example of the world of meanings, created by people to facilitate their daily lives. As with all such creations, they are made in context, through interactions with others. Thus our taken-for-granted worlds are both individual and shared. And they change too, as new types are required, and reallocations seem desirable (what was a 'safe street' to walk in at night no longer is).

Such basing of behaviour on inter-personally created stereotypes of people, places and institutions is typical of all aspects of life and not just the day-to-day living in the local environment – what we usually study under the title of social geography. Ley (1980) has illustrated this in his work on The Elector's Action Movement in the City of Vancouver, but this sort of study is rare. And yet, if we believe that the contexts for the local community are the nation-state and the global economy, then clearly we need to understand the taken-for-granted worlds of the powerful decision-makers at those levels also. They, too, operate on stereotypes, which condition their attitudes towards people and places, and hence their decisions regarding, for example, the location of jobs. To the managers of British capitalism, some places are attractive venues for investment (e.g. Swindon) whereas others (e.g. Sheffield) possess a very negative stereotype. We infer that from their decisions, and we gain some information from their statements and interviews. But we know very little about their taken-for-granted worlds, how they are produced, reproduced, and restructured. Instead, we use our own, for as human geographers we too operate through stereotypes to which we allocate our subjects. To the extent that there is dissonance between our stereotypes and theirs, so we are failing to understand them. And so we portray *our* geographies (those we care to believe) rather than *theirs*, and yet it is the latter that are important, not ours.

Much of this section will be interpreted, as it was intended, as a case for an ethnographically based description of particular actions. If places matter, because they are critical contexts, then we must understand those places, as they – as well as the structures – are interpreted by the people within them.

Following on from this, a second conclusion that is likely to be drawn is that work on large-scale description of the world of experiences has a lower priority, particularly work based on the researchers' definitions rather than the actors'. This is not necessarily so (and not just because it is that sort of work on which my reputation as an empirical researcher has been based). There are several reasons for this. First there are issues of scale, both spatial and numerical. Many insights into global and national worlds are only attainable through external data sets – though care must be taken in their use, as Ramphal (1985) brilliantly illustrates. The same is true for large population aggregations; only an outsider's view of the residential areas of Mexico City can provide a general oversight, even though for full understanding it must be supplemented by insiders' views. Secondly, there is the point that we are not the only outsiders imposing our stereotypes on the world. Human geographers occupy no privileged position and many others, including those with power to change the world, are using the same data; research interpretation of those data is vital, therefore. Thirdly, however desirable ethnographic studies are, only a very small number of case studies can ever be collected, and so they must necessarily be supplemented by the synoptic studies, with the two mutually reinforcing each other. Fourthly, and very importantly, ethnographies can, very obviously, deal only with the present. And yet, as the structuration concept makes clear, the present is just a moment in the continual reproduction and restructuring of people, societies, places, and structures. The past is being carried forward, creating a palimpsest, not only in the landscape – which is the product of successive layers of investment and disinvestment – but also in people's minds and in their private and collective stores of knowledge (what Popper calls World 3). Thus to understand change we need to be able to interpret those palimpsests. Finally, Pickles (1985) has argued that:

> The interpreter of an event, piecing together the picture of what happened may well come to know that event better than those involved (but in a special way only).

Such outsiders' interpretations are of value, he argues, because:

> if we did accept literally the word of those who brought ideas to light in the first place then we may as well not re-think those issues. Indeed, if we merely accept the claims of previous geographers at face value . . . we run into the problem . . . of how we are then to avoid passively accepting the status quo, and thereby perpetrating existing ideology . . . In re-thinking arguments we seek precisely to see them in a new light; to problematize them, in order to move beyond them. (p. 173)

The insiders live in their worlds. If the task of the scientist is to liberate them from the constraints there (see chapter 6), then they must put those insiders' worlds into their broader context. How to do that is the topic of the next section.

STORY-TELLING

> [History is] an argued invitation to imagine the intricacies and the coherence of a condition of human circumstance which has not survived.
>
> M. Oakeshott, 1983, p. 58

Hermeneutics is a term that has recently entered the geographer's vocabulary. Its original meaning related to the interpretation of the Bible and other theological texts in order to find the spiritual truths deposited there. It is now more generally used to describe the general skill of interpretation from a text, identifying the meanings that it expresses: it is particularly associated with the work of Wilhelm Dilthey (Rose, 1981).

'All history', according to Guelke (1982, p. 53), citing Collingwood, 'is the history of thought', so that historical analysis is 'the re-enactment of the thought in actions'. One cannot study history directly, but only through the texts − the relicts of the past. For most historians, those texts are written words, the documentary sources on which they work: most historical geographers follow them. But there are many other texts, selective outcomes of events: for the study of recent history, there are the memories of participants; for the study of prehistory (i.e. of pre-literate societies), there are the artifacts painstakingly discovered by archaeologists; for the study of any period there is the landscape and there is the representation of society in a range of art forms (literature, painting, music, even maps etc.). All of these are texts, which tell us something about their creators and their worlds, and so analysis of those texts allows us to penetrate people and worlds to which we have no direct access. Such analysis is clearly a form of hermeneutics. In much reporting of humanistic work, involving direct contact with subjects, a double hermeneutic is undertaken: the respondents interpret their lifeworlds to the researchers, who in return interpret them to a wider audience, so that the mutual understanding, which is the hermeneutic goal, is obtained through two filters. (There is usually a third, too, since the research is only reported indirectly, through the written word. Thus the readers of the research reports are interpreting them through their own cultural lenses; to some extent, they read what they want to read, or at least ignore that which they do not want to read.) In the search for mutual understanding through texts, there is a triple hermeneutic, since the researcher has to interpret the link between the text and the creators.

A major area of hermeneutical analysis is the study of literature, which involves studying the texts to obtain insights into the nature of the world being portrayed. This is demonstrated, for example, in the work of Raymond Williams, whose *Culture and Society 1780-1950* (1958) uses literature to trace changing British culture during that period, and whose *The Country and the City* (1973) examines the urban:rural division of society through the same medium. Human geographers have recently moved to the edge of this field, as in Douglas Pocock's (1981) collection on *Humanistic Geography and Literature: Essays on the Experience of Place.*

His subtitle indicates that the concern is not with either the 'geography behind literature' (with all the potential for collapsing into environmental determinism) or 'geography in literature' (reconstructing the places discussed); rather it is with 'exploring the nature and aspects of environmental experience as part of the human condition' (p. 15), with literature offering the perceptive insights of the author: 'The writer . . . articulates our own inarticulations about place, our fellow men and about ourselves, providing thereby a basis for a new awareness, a new consciousness.'

Those who use works of literature as texts for gaining insights into other times and/or places do not come to the individual works *in vacuo*. They seek to set them in context, particularly the context in which they were written. This means that they need background information on authors and their lifeworlds, in order to appreciate what the purpose of the writing was. Thus, for example, R.A. Butler's introduction to the Penguin edition of Disraeli's *Sybil* (1980) refers to his novels being 'circumscribed by what he knew and liked best' and stressed that 'we must not allow ourselves to forget . . . the uniquely political aims of the author' (p. 10). In writing that novel, Disraeli was providing his views on the current social distress that had provoked the Chartist disturbances, and thus gives us insights not only to those disturbances but also to his interpretations of them – he is portraying his world of experience through a particular art form. Similarly, Trollope's Palliser novels were written by a failed politician during the period of the second Reform Bill, and in his autobiography he stated (as cited in Sutherland's introduction to the Penguin edition of *Phineas Finn*, 1972, p. 12): 'As I was debarred from expressing my opinions in the House of Commons, I took this method [the Palliser novels] of declaring myself.'

Human geographers, as already noted, have been using literature to get insights into the 'sense of place', people's affinitive links with their physical and social milieux, in statements such as:

> . . . There was a sound of tea being cleared away in a cottage just near us. And suddenly with a burst the bells of Highworth church rang out for Evening Service. As though called by the bells, the late sun burst out and bathed the varied roofs with gold and scooped itself into the uneven panes of old windows. Sun and stone and old brick and garden flowers and church bells. That was Sunday evening in Highworth. That was England. (Sir John Betjeman, 1978, p. 210)

Clearly a full appreciation of this needs some understanding of the author, setting him and it in context. But there is more to literature as a source than that, as recent structuralist debates have suggested. There are deeper meanings, as Williams argues in *The Country and the City*. He notes the continued appearance of the dichotomy between 'the idea of pastoral innocence . . . [and] of the city as a civilising agency', and asks 'what kinds of experience do the ideas appear to interpret, and why do certain forms occur or recur at this period or at that?' (p. 290). Deep-seated concepts (essences?) seem to reappear in a range of historical realities.

To explore those deeper structures, which cannot be directly apprehended, one needs theories, and so people who use texts must do so in a theoretical context, not, as Guelke would argue for idealism and as phenomenologists would argue with regard to 'bracketing', in a theoretical vacuum. This need for theory is made clear in Williams's work. In *Culture and Society* he uses texts in the context of a theory of societal change built around industry, democracy, and art:

> The history of the idea of culture is a record of our reactions, in thought and feeling, to the changed conditions of our common life. Our meaning of culture is a response to the events which our meanings of industry and democracy most evidently define. But the conditions were created and have been modified by men. Records of the events lie elsewhere, in our general history. The history of the idea of culture is a record of our meanings and our definitions, but these, in turn, are only to be understood within the context of our actions. (p. 295)

That context is both proximate and structural, with the one linked to the other, as made clear in *The Country and the City*.

> The industrial-agricultural balance, in all its physical forms of town-country relations, is the product, however mediated, of a set of decisions about capital investment made by the minority which controls capital and which determines its use by calculations of profit . . . it is not only that the specific histories of country and city, and of their immediate interrelations, have been determined, in Britain, by capitalism. It is that the total character of what we know as modern society has been similarly determined. (p. 295)

The texts illuminate not only the worlds of experience and events, but also the interpretations of the world of mechanisms. By using the texts in a theoretical context, we gain insights not only to lifeworlds but also to the creation and recreation of those lifeworlds by human agents acting within structures.

Literature, basically fiction, provides only one set of texts. Indeed non-fictional literature (including geography) is similarly a set of texts. Understanding why a geographer wrote a certain piece, or conducted a research programme in a particular way, similarly requires an appreciation of the context, of the geographer's interpretations of the three worlds – empirical, actual and real. With that understanding, we are better able to use the text to inform the wider project, of understanding the constitution of society.

Human geography, then, is the interpretation and creation of texts. The texts interpreted cover the full range of sources – oral, written, landscape, artifacts etc. Those created are usually written, statements which bring together the texts in an ordered arrangement. (For human geographers, maps and landscapes are the main alternatives to written texts.) Thus the geographer takes the texts of others, including other geographers, and weaves their contents into a viable synthesis; stories are assimilated into others. Those syntheses are theories, constructions of the 'reality' being studied. The fundamental feature of such a theory is that it should cohere, that the parts link together. As new material is obtained (more texts examined) so the theory is tested; does it continue to cohere? If the parts do not fit, the theory is invalid – it is telling a story which does not ring true with the evidence, and it must be modified accordingly. To some considerable extent, therefore, the human geographer is acting in the critical rationalist way promoted by Sir Karl Popper, creating conjectures and seeking to refute them through 'testing' them against reality – and there is no necessity for this to be positivist; the goal is a coherent understanding, not a set of laws. Peter Hall's (1980) study of *Great Planning Disasters* provides a good example of this, seeking to appreciate how and why certain types of decisions were made – to provide lessons as a learning experience, because the past is understood, but not narrow blueprints for future action.

In setting out to build up a story, researchers do not begin without preconceptions. As Collingwood (1965, p. 39) expresses it, the historian begins with the skeleton of a theory, built up, no doubt, from a combination of intuition, other historians' texts (perhaps relating to other times and places, and so offering analogies only), and general attitude to the subject matter, reflecting the process of socialization as an historian in a particular academic context: in other words, there is 'some working hypothesis as to the things especially worth noticing' (p. 39). But no more. The historian is seeking to understand a *situation*, and is responding to 'an argued invitation to imagine the intricacies and the coherence of a condition of human circumstance which has not survived' (Oakeshott, 1983, p. 58). All of this involves the reconstruction of empirical and actual worlds, the past (including the immediate past) as it was experienced and acted in. To Guelke (1982, p. 30) nothing more is needed:

> The explanation of an action is complete when the agent's goal and theoretical understanding of his situation have been discovered. It is not necessary to investigate the grounds on which a particular theory might be entertained, because they are irrelevant to understanding an action that might be related to it ... *One must discover what he believed*, not why he believed it [original emphases].

Thus in studying English voting patterns, for example, it is sufficient for me to know that some people believed it to be in their and the national interest to vote Conservative, and no more. The existence of the Conservative Party is not problematic, it seems, nor is the reason why people are convinced they should vote for it. Apart from anything else, this appears to be a remarkably ahistorical approach to understanding for an historical geographer. Without appreciating either the origin of the choice set (the various parties) or the interpretations of it, through the processes of political socialization in home, neighbourhood, workplace and other milieux, one is unable to achieve that which Pickles (1985) advocates – a better knowledge of what happened than that possessed by the actors. And if, as with much human geography, the actors are still alive and active, then the absence of that wider interpretation will, as Pickles argues, ensure the maintenance of the status quo. Geographers do not have to be committed to a particular form of social change, but they should be committed to a broadening of horizons, to creating texts which are more than simple aggregates of others, in order to advance people's self-knowledge and ability to construct change.

Empirical work is insufficient, therefore: it must be set in a theoretical context, with which it coheres, if it is to do more than reproduce society. Thus human geographers must interpret texts (the worlds of experience and events – either directly or through the agents concerned) in the context of their interpretations of structures; figure 4.2 is as relevant to the practice of human geography as it is to the practice of daily life. Just as the individual is continually interpreting the world, in order to survive in it, in the context of a world-view, a theory of society (however poorly formed), so should the academic interpreter of that individual's actions. The individual simplifies the world in order to live in it, creating stereotypes and making the world fit them, and only occasionally modifying the stereotypes. Too often the academic does the same, creating models that constrain and limit rather than enable understanding; this is the lesson of Kuhn's paradigm model of scientific progress, with the continued unpreparedness of researchers to face up to the mounting body of anomalies which threaten the viability of their models. The researcher as researcher must be more open to attacks on the stereotypes than the researcher as citizen, otherwise the goal of the broader perspective will be submerged under prejudice and self-interest. Nearly 50 years ago, Davie (1938, p. 136) criticized the adherents to Burgess's zonal model of the city for their apparent attitude that 'the hypothesis must be maintained whatever the fact may show'; 39 years later, Guelke (1977, p. 48) accused urban geographers of a similar crime, with regard to tests of central place theory, that their rule was 'one counts one's hits but not one's misses'. Theory and empirical findings must constantly interact, with the latter enriching the former, and the former guiding the latter.

The term guiding is crucial in the previous sentence, because the theory to be used in this context is not the type of theory that most human geographers identify with – that is, the type used in positivist science. In the latter, theory is the basis for deductions and hypotheses that can be tested empirically: a good theory is one that can successfully predict an event. This is not so in the type of theory associated with the realist approach to science, advocated here. That is a theory of the real world, of the structure in which the mechanisms operate. But those mechanisms are not deterministic, because they create a situation which enables choices to be made. Within that enabling framework, the contingently related circumstances (figure 4.3) may constrain, even curtail, the choice set. Those circumstances are not fixed, of course, because they are part of the superstructure that is constantly being reproduced and altered. Hence predictions at the level of events and experiences are, at best, very tentative because the circumstances in which the theory is being put into operation are always changing.

If the theory does not provide predictions to be tested in empirical work, what does it provide? To use an excellent geographical metaphor, it provides a map, an outline of the choices that are available. To take an example, a private steel company in Sheffield has decided to close its works, and create thousands of redundancies. Those directly affected – the workers – contest that decision, through their trades unions, and seek support from the representatives of the many others indirectly affected – the City Council. Together, this alliance of local interests petitions the central state to intervene and prevent the closure. The options open to the state apparatus are several. The petition can be rejected, and nothing done at all – other than policies already in place – on the grounds that there is clearly over-capacity in the steel industry and closure is necessary; the workers must accept this as part of the inevitable process of change involved in industrial restructuring during a recession. Or, the petition can be rejected but the state recognizes that it will cause many problems for the local labour force, so it offers extra money and facilities for retraining, for the establishment of new businesses and, if necessary, moves to other parts of the country.

The two options canvassed so far have no direct state involvement in the company itself. Other options do, and the state may decide to prevent the closure by subsidizing the company. There are several ways in which it can do this. One is to subsidize attempts to increase productivity and promote profitability through investment in new technology; another is to enhance profits by subsidizing labour costs.

The map outlines the options, therefore; it does not predict which will be selected. It does provide other keys to understanding the choice, however. As outlined in chapter 3, the state's roles are to promote and legitimate the mode of production and to maintain social cohesion, and it takes decisions in this context. Failure to promote accumulation can stimulate an accumulation crisis; failure to maintain legitimation could generate a legitimacy crisis; failure to maintain social cohesion could produce a rationality crisis. The individuals operating the relevant state apparatus must interpret the situation, decide which, if any, crises are likely, and act accordingly. As always, those individuals are drawing on their interpretations of the mechanisms and empirical situations, in their particular contexts. They may be following a clearly laid out route through the map; more likely they are reacting to events.

The theory that informs our study of these events is one which provides us with the map of the terrain occupied by the actors involved. It shows us the routes, and helps us to understand which is chosen – provided that we appreciate the context in which that choice is made. It does not provide the base for predictions, as Andrew Sayer (1982, p. 85) makes clear, because: (1) we cannot create experimental situations and hold constant all the contingent conditions, so we can apply the theory in a context, but not test it; (2) the theory shows many routes forward, and explains why they exist, but can, at best, only tentatively suggest which might be chosen (probably in conditions of fully informed, rational decision-making that never apply to people); and (3) we could only predict if we knew the future, which we do not because we do not know how knowledge will develop, and one problem of making predictions is that they can be negated – people can react negatively to them. All we can do is evaluate the theory; has it helped us to understand events and to chart a future? If not, clearly the theory is incomplete. One possible reason for this, of course, may be that in order to solve the problems that they face the actors create new routes, which is of course what has happened throughout the history of capitalism (e.g. the creation of joint stock companies and then of multinational corporations). Our theories, then, must be constantly updated if they are to be valid guides to the decision-making map.

Without theory even description is impossible, for if we have no conceptual equipment we have no medium with which to convey our descriptions – to create texts. All research is theory-laden, however unconsciously (and however much some people may feel that they need no theory: Taylor, 1983). The sort of theory that we need is that outlined in chapter 3, providing us with the evolving, necessary framework for the appreciation of that which we experience and do, and that which others experience and do.

Within the context of that theory, which is itself a text on which we draw, we produce our texts, our appreciations of the actual and the empirical. In doing this, we produce a story that coheres, empirically and theoretically, and which as a consequence enhances our appreciation of what people do, how, and why: the what is empirical description; the how is the actors' logic for the events; and the why puts their logic into its theoretical context.

This approach to research enables us to be eclectic, to draw on texts produced by researchers operating within other approaches. What they provide us with may be limited in its utility, but nevertheless of some value. Thus an empiricist description of who voted what, where, gives us material that can inform an attempt to appreciate the workings of liberal democracy, and positivist attempts to predict votes from ideological positions similarly tell us who accepts certain ideologies, where. We do not accept the positivist tenet that we are being provided with some universal guide to predicting votes in the future, but we do get empirical insights that can, with care, illuminate our attempt to develop a theoretically informed position, using theory in its realist definition. Similarly, idealist accounts of political party activity – their rationale for campaigning as they did – provide us with textual information that can be incorporated in the more general programme, and which may inform the development of theory as well as appreciation of the actual and real.

Clearly research adopting the realist approach would advance more rapidly if all were working within the same framework. Other approaches slow down realist progress, but they are not entirely divorced from it. We interpret texts, produced other than for research purposes, within our framework, fully realizing that those who produced them were not necessarily in agreement with our theory, and we are able to do this because we appreciate – theoretically and empirically – the contexts within which they were produced. Exactly the same is true of research texts: indeed, they should be easier to use because we are more likely to appreciate the conditions in which they were created – though few academics have accepted Anne Buttimer's (1974) invitation to lay bare their own 'values, prejudices, and hopes' (see, however, Buttimer, 1984, and Billinge, Gregory and Martin, 1984).

IN SUMMARY

Realist research involves developing an understanding of the world of experiences – what we see there, and what others see – through the world of events – what people did and do, and their reasons for it – and into the world of mechanisms – the structures which both enable and constrain action. It involves story-telling, putting together texts from other texts, so that they cohere and provide us with a framework which allows us not only to appreciate the world that we live in but also to act constructively in it.

Debate over philosophy within human geography in recent years has produced a variety of reactions. One of the commonest is the pluralistic ('let every flower bloom') attitude, as expressed by Golledge (1982, p. 15):

> the discipline should not be dominated by one mode of thought, and . . . perhaps it is the existence of such 'divisions' in geography that has the potential to make it one of the more imaginative and exciting disciplines of the current decade.

The present chapter could be considered another contribution towards that pluralistic orthodoxy. It is not, however. The philosophy advanced here is avowedly realist, on the grounds that full understanding necessitates an integration of the real, the actual and the empirical. Within that realist approach, understanding requires empirical research. Clearly, investigations of both the empirical and the actual world, conducted within the realist framework, are likely to produce the greatest returns, but the texts produced according to other philosophical approaches can provide valuable inputs, if treated sensitively through an appreciation of the context of the research traditions. (To return to the example of the closure of a Sheffield steel plant, an empiricist discussion of the debates and the outcome would indicate which route was selected, even if the other options were not discussed and the reasons for the choice went unconsidered. The empiricist discussion would be partial, because it would lack any reference to the closure within the general context of industrial and geographical restructuring, but its text would be of value to a properly structured realist account.) Indeed, hermeneutics are relevant to all philosophies, for the major purpose of any research report is to communicate its findings to others.

As individuals, we live in a world of texts, and our interpretations of them condition our behaviour. As human geographers, we work in exactly the same way: we have our disciplinary texts – theories, models, research reports etc. – which we interpret in order to guide our practice of human geography. That practice involves the study of other texts in the context of our own. We use whatever information we can obtain to tell convincing, coherent stories about the stories our sources tell us, in word, deed, picture, map, landscape, or whatever. And so our stories are never complete; as we learn more from the texts, so we enhance our narrative.

There are no protocols for such an enterprise, although texts not based on chaotic conceptions are clearly preferable, and no tests of the validity of the outcome; the evaluation of a story lies in whether it helps us to understand, and so to plan our route forward. And not every text we create is going even to attempt the whole story, for we build it up slowly through separate pieces of work. Individual texts may not disclose their position in the full story, perhaps because the latter has yet to be formulated. But they are building-blocks, sentences and clauses to be put together later into a coherent narrative. One day, perhaps, we will tell the full story; until then, we will go on writing small portions of it.

5

Positivism,
Science and Quantification

Like a logical positivist, he was thinking in the shallows of his mind,
while in the depths the great Event of light and sound timelessly
unfolded . . . the Logical Positivist, absurd but indispensable, trying
to explain, in a language incommensurable with the facts, what it
was all about.

Aldous Huxley, *Island*, 1976

Geographical story-telling, as outlined in the previous chapter, incorporates
theoretical studies of mechanisms with empirical studies of actors, events
and outcomes into a realist scientific account. On its own, theoretical study
is of little value if it does not illuminate the empirical world for, I argue,
people will be unconvinced by the theory as a guide to practice if it does not
help them to appreciate the worlds of experience and events. My particular
concern in this chapter is with aspects of the conduct of empirical inquiry.

Much of the contestation over philosophical issues among human
geographers in recent years has been concerned with the merits and demerits
of empiricism in general and positivism in particular. The latter was
incorporated into the discipline from the mid-1950s on, though not fully
enunciated by a geographer until David Harvey's (1969) statement in
Explanation in Geography. Positivism was introduced to social science
through sociology, though it first gained major status within economics.
The 'positivistic attitude', according to Giddens (1974, pp. 3–4), comprises
three basic propositions:

1 The methodological procedures of natural science can be directly
  adapted to sociology, so that phenomena relating to human values
  and actions are treated as objects to be studied in the same way as
  those investigated in the natural sciences;
2 The outcomes of such applications to sociology will be the same as
  those in the natural sciences, statements of a law-like character; and

3   Those law-like statements can be used as instruments in the conduct
of practical policy.

If we replace sociology by human geography, we get the fundamental
features of the positivist approach to our discipline. Keat (1981) extends
this case, by pointing out that adherence to positivism – or 'the positivist
conception of science' – involves three doctrines: (1) scientism, the claim
that the positivist method is the only true route to knowledge; (2) scientific
politics, the claim that positivism provides the rational solution to
problems; and (3) value-freedom, the claim that positivist research is
'objective', 'neutral', and 'value-free'.

The critique of positivism in human geography, as in other social sciences,
has concentrated on the points raised by Giddens and by Keat. In particular,
it has focused on the pursuit of general laws. Thus Wirth (1984, p. 73) has
claimed that: 'Thirty years of every-day research routine in the empirical
social sciences have shown that a strict orientation towards the *general*,
towards *laws* and *comprehensive theories* lead into a cul-de-sac.' Derek
Gregory (1980) has been particularly critical of the instrumentalist position
that underlies positivism, as indicated by the quotation from Bennett (1974,
p. 172) that 'whilst the real world certainly does *not* behave as a low order
TF or ARMA model, it *can* be treated as such.' Gregory argues that the
uncritical application of such models ensures reproduction of the status
quo, so that there 'is as much danger in its succeeding as there is in its
disappearing altogether' (p. 341). Elsewhere (in Baker and Gregory, 1984)
he notes that this was but part of a general functionalist movement 'which
was unequivocally opposed to historical geography on any other terms
other than as a 'testing ground' for general propositions which rejected any
suggestion of space-time specificity' (p. 183). And finally, some writers
(notably Sack, 1974) have pointed out the logical impossibility of a separate
discipline of human geography (spatial science) based on distance as the
fundamental variable.

Responses to such critiques have been several and varied, many of them
based on misconceptions of the nature of positivism and the arguments
against it. Positivism has been equated by many with, for example, both
quantification and the hypothetico-deductive method (known to many as
'*the* scientific method': see Abler, Adams and Gould, 1970). And yet neither
is central to either positivism or the critique. Bennett (1981), for example,
has argued strongly against the bracketing of quantitative geography with
positivism, and rightly so, but fails to provide an equally spirited defence
against the more fundamental charges of law-seeking instrumentalism.
(Central to the latter is the argument that the application of positivist laws

reproduces the existing structure of society and cannot promote 'real' change; see chapter 6.) He admits, for example, that much of the work in the tradition has 'consciously sought objectivity, technical expertise, spatial description, and *inductive approaches directed towards revealing universal laws which might be used to plan and control*' [p. 13; emphasis added]. Golledge, too, in his defence of behavioural geography (1980, p. 16) argues that: 'by uncovering individual behavior patterns and improving one's understanding of the reasons for such patterns, the researcher could aggregate populations into meaningful and significant sub-groups *such that significant generalizations could be made about those particular sub-groups*' [emphasis added].

To some extent, clearly, the critique and the defence of positivism are being argued on different grounds. In particular, this has led to a widespread belief that quantification and hypothesis-testing are invalid *methods* because they have been used for the development of instrumentalist *ends*. To confuse means with ends is to do an injustice to the means in many cases, as will be argued here.

HYPOTHESIS-TESTING AND THE SEARCH FOR KNOWLEDGE

The methodology associated by most human geographers with positivism is the organized procedure of speculation and verification/falsification, which many of them know as '*the* scientific method'. (Some claim that this association comes from Harvey, 1969, but he referred to that procedure – p. 34 – as 'An alternative route to scientific explanation'.) Amedeo and Golledge (1975), for example, call it the process of *scientific reasoning* whose purpose is the search for spatial laws (p. 23), and King and Golledge (1978) have a section entitled 'scientific method' which presents the 'scientific approach' as having four stages: observation, theory-development, test of the theory empirically, modification of the theory in the light of the test results (pp. 6-7).

According to the usual representations of this procedure (drawn from a diagram produced by Harvey, 1969, p. 34 – figure 5.1 – and repeated many times) one begins with a problem, and then seeks to discover what we already know that can help in its solution, either directly or indirectly (e.g. through the use of analogues). The result is a model of the problem area, a generalized statement in diagrammatic, symbolic or verbal form. This is a map against which reality is to be compared, through hypothesis-testing. The hypothesis is deduced from the model as an unambiguous statement of what should be observed empirically, phrased in such a way that it can readily be validated. An 'experiment' is then conducted to test the hypothesis. If it proves valid then one has gained positive knowledge; if not, then the outcome is negative knowledge. (Negative knowledge is frequently derided, but if properly obtained – and therefore valid – it is as useful as

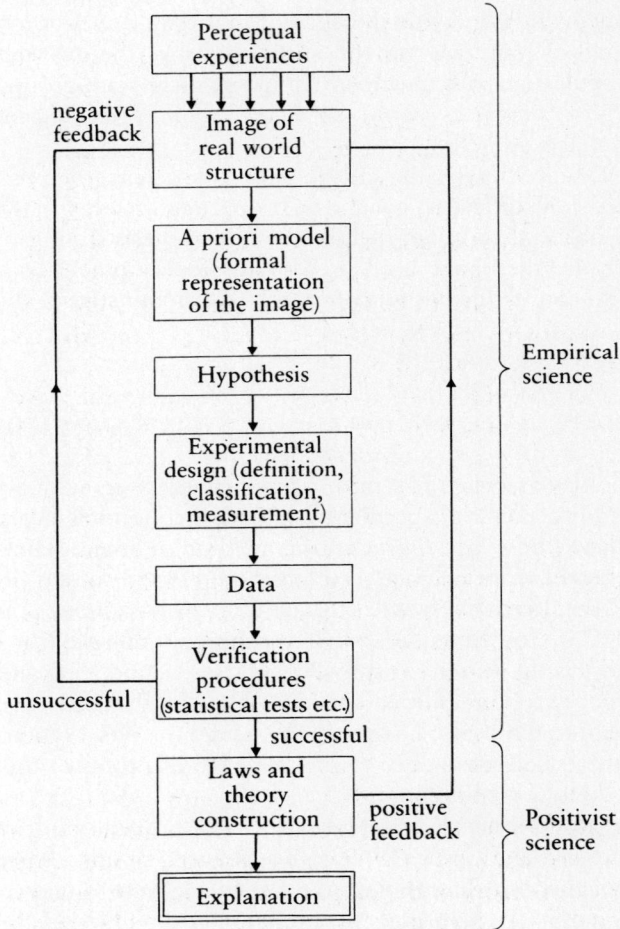

```
                    ┌──────────────┐
                    │  Perceptual  │
                    │  experiences │
                    └──────────────┘
                      ↓ ↓ ↓ ↓ ↓
   negative         ┌──────────────┐
   feedback         │   Image of   │              ⎫
                    │  real world  │              │
                    │  structure   │              │
                    └──────────────┘              │
                           ↓                      │
                    ┌──────────────┐              │
                    │ A priori model│             │
                    │  (formal     │              │
                    │ representation│             │  Empirical
                    │ of the image)│              │  science
                    └──────────────┘              │
                           ↓                      │
                    ┌──────────────┐              │
                    │  Hypothesis  │              │
                    └──────────────┘              │
                           ↓                      │
                    ┌──────────────┐              │
                    │ Experimental │              │
                    │design (definition,│         │
                    │ classification,│            │
                    │ measurement) │              │
                    └──────────────┘              │
                           ↓                      │
                    ┌──────────────┐              │
                    │     Data     │              │
                    └──────────────┘              │
                           ↓                      │
                    ┌──────────────┐              │
                    │ Verification │              │
                    │  procedures  │              │
   unsuccessful     │(statistical tests etc.)│    ⎭
                    └──────────────┘
                           │ successful
                    ┌──────────────┐              ⎫
                    │  Laws and    │              │
                    │   theory     │   positive   │  Positivist
                    │ construction │   feedback   │  science
                    └──────────────┘              │
                           ↓                      │
                    ┌──────────────┐              │
                    │ Explanation  │              │
                    └──────────────┘              ⎭
```

Figure 5.1   Scientific procedures and positivist scientific method (modified, with permission, from figure 4.3 of Harvey, 1969, p. 34)

positive knowledge; it is certainly not useless to discover that water does not flow uphill naturally – though one might wonder why such a phenomenon was postulated in the first place).

Operationalizing this procedure has produced some problems for human geographers. The term 'experiment' suggests a laboratory, in which all other influences (the 'contingently related conditions' of figure 4.3) are held constant, so that the test of the hypothesis is not in any way contaminated. Such experiments are only possible in human geography (as in much physical geography too) in artificial situations that are so divorced from what they are supposed to represent that the results are not transferable. And so the 'experiments' must be conducted in 'contaminated situations', with as much care as possible so that as many of the contingently related conditions are neutralized as can be. As a consequence, the results of the tests are usually equivocal, and it is uncertain whether the hypothesis has been verified – how does one interpret an R^2 of 0.3? Only tentative conclusions can be drawn, to be fed back into further rounds of organised speculation, improving the hypothesis to take account of circumstances ignored previously (as outlined by Hay, 1979).

As set out so far, this rigorous scientific procedure raises no problems with regard to the story-telling approach advocated in chapter 4. Indeed, it differs only in its terminology from the procedures of idealism proposed by Collingwood (1965, see p. 76), and the term hypothesis is not noticeably absent from some historians' vocabularies. As argued in that chapter, the goal of story-telling is to develop a description, of the empirical or actual world being studied, which coheres. One has the skeleton of a description, evaluates a set of texts, fleshes out the skeleton in certain areas, returns to the texts, and so on. If done rigorously and honestly, there is no difference between what has been described here and what was advocated there.

The difficulty with the positivistic conception of science is not so much with how it obtains its material – how it addresses the texts – as what it does with it. This is illustrated by the adaptation of Harvey's well-used diagram in figure 5.1. The final two stages of the sequence shown there are 'Laws and theory construction' and 'Explanation'. The results of the hypothesis-tests not only feed back into the 'image of the real world structure' – the scientist's coherent theory of the subject matter – but also feed forward into the development of scientific laws, defined by Harvey (who drew substantially on Braithwaite's *Scientific Explanation*, 1960) as: 'A scientific law may be interpreted most rigidly as a generalisation which is empirically universally true, and one which is also an integral part of a theoretical system in which we have supreme confidence' (p. 105). These laws are then used as the raw materials for *explanation*, in which particular events are accounted for as examples of a general law.

It is these final two stages that are the focus of the criticism of positivism, not the entire sequence; the attack is not on the rigorous evaluation of speculations but on the belief that the answers to those speculations lie in general laws, a belief which implies that, ultimately, all behaviour is determined by external factors. In the social sciences, this treats the human as an object acting according to predetermined rules, which runs counter to the concept of the human agent in realist approaches. (Note that explanation takes this particular form in the positivist philosophy only. In realism, as Sayer (1984) argues, one can produce an explanation, a statement of what caused an event, but one does not assume that this applies to other events. Whereas in positivism a prime characteristic of explanation is replication, in realism it is corroboration. Positivists explain with reference to general laws; realists explain with reference to the particular case.)

The central stages in the sequence (figure 5.1) involve the rigorous evaluation of hypotheses, but we must identify two types of hypothesis:

1   The unrestricted hypothesis, relevant to situations beyond that in which the test is taking place, because the test situation is considered representative of a large number of such situations; and
2   The restricted hypothesis, relevant only to the context within which it is being tested.

Unrestricted hypotheses are irrelevant to the approach to human geography developed here, because the transferability of findings is an unacceptable axiom; but restricted hypotheses are acceptable. They do not necessarily have to be tested in the way Harvey lays out, however: classification, measurement, statistical tests etc. may be useful tools in the descriptive task for which the hypothesis-testing is being undertaken, but they are not the sole tools, and rigorous evaluation of any type of text provides a perfectly acceptable means of testing.

The empirical world is the outcome of decisions and the interpreted context for future decisions. The nature of that interpretation is crucial, therefore, so how do the actors involved describe the world that they experience? Does their description tally with that of others? If you want to understand what they did, how do you set about discovering how exactly they acted? The testing of restricted hypotheses can be a substantial aid in answering some of these questions, most of which will be posed in the context of a theory, the story that is being constructed by the researcher. To return to the example of the steel company closure again (see p. 78), the theory of available routes provides a context for studying what actually happened, and why. The theory offers no account of the contingently related conditions, and so cannot predict which route would be taken. Of course, given that we are dealing with human agents, even if all the conditions were known prediction would still be impossible, since it would assume general laws. But strongly argued reasons for one route would be possible, and they could be evaluated. The researcher, in building up a story, will be aware of the contingent conditions, and so can suggest which route is likely to be taken. As a hypothesis, this guides the next stage of the research, directing the way in which the texts are addressed.

Such a use of hypothesis-testing is especially profitable in the sort of study, typical of much human geography, which has material on the outcomes but not the events itself. This is, of course, true of almost all historical geography and also of much 'contemporary historical geography'. Further, in many cases the human geographer has information on a large number of outcomes, but not at the individual level. In creating a coherent and convincing story, therefore, it is necessary to deduce events from outcomes. Thus, for example, in a study of the 1983 general election in England I anticipated that a shift away from Conservative and towards Labour would be most marked in the constituencies with the highest levels of unemployment (Johnston, 1984b). This anticipation arose from my understanding of: (a) the general processes of political socialization in Britain, whereby a large majority of middle-class voters have accepted the Conservative, free-market ideology and a majority of working-class voters have accepted the Labour, state-ownership ideology; (b) the erosion of Labour working-class support by the extension of the Conservative ideology in 1979; (c) the large rise in unemployment between 1979 and 1983, which affected the constituencies differentially; and (d) the Labour claim in 1983 that its proposed policies would 'put Britain back to work' whereas those of the Conservative party suggested a continuation of the policies that opponents claimed had contributed substantially to the rise in unemployment. Thus through an appreciation of the contingently related conditions, I could suggest the outcome indicated above, a hypothesis that could be rigorously tested. (There were other contingent conditions at the time of the election, of course: Johnston, 1985a.) If it proved valid, then I had gained some understanding of how people acted in 1983, and the story that I was building up was bolstered accordingly; if it proved invalid, then clearly either I did not understand the contingently related conditions, or I was a bad judge of how people, in the aggregate, would respond to them.

The hypothesis proved valid, thereby advancing my understanding of the 1983 election. But the results of my test did not produce a general law of electoral behaviour, either applicable to other places or relevant to other times in the same place. The contingent conditions will never be repeated in exactly that form and, with regard to England in the future, the results of the 1983 election and political changes since will alter the milieux in which voters act. I could study the next election using the same hypothesis, acting as if the contingently related conditions remained constant (shades of instrumentalism), but it is surely better always to study the operative conditions and hypothesize accordingly. To operate from an historical analogy alone, however close in time, is to deny the constant dialectic of change which is the case of the realist position. Some of the contingently related conditions change more rapidly than others, of course – the events surrounding an election are particular to it, but the political culture(s) in which it is conducted may be well-established in places (Johnston, 1985c) – and these various rates of change are reflected in the story that I put together and the empirical questions that I ask.

All of this testing of restricted hypotheses takes place at the empirical and the actual levels only. It is a part of the description, the story-telling about the actions, and their outcomes, that occur in the context of the contingently related conditions. There are no tests of the theory of the mechanisms because, as stressed before, these must be seen as enabling as well as constraining choice. We cannot explain outcomes in the positivist sense with reference to mechanisms, we can only interpret them as the products of those mechanisms set in operation in particular, interpreted, contexts. Nor can we explain in the positivist sense with reference to contingently related conditions. But we can explain in a realist sense, we can say that people interpreted their situation in this particular way, and acted accordingly; they may never pass that way again.

POSITIVISM AS SCIENCE AND POSITIVISM IN ACTION

There is one counter to the final statement of the previous sentence. The world in which people live is structured by them – though not, of course, in conditions of their own choosing. As indicated earlier (p. 77), people react to the complexity and difficulty of the world and of taking decisions in it by simplifying it, by reducing it to a set of stereotypes and then acting habitually; they make unconsidered choices in a taken-for-granted world rather than taking decisions in a world of continually changing, contingently related conditions. In this sense their actions are foreseeable, and one should be able to predict their actions according to the full positivist sequence in figure 5.1; general laws of behaviour are possible, because that is how people choose to act. To evaluate this claim, I turn to a discussion of generality, uniqueness and singularity.

Generality, Uniqueness and Singularity

The concern with generality, with the focus on laws that are invariant over time and space, is the nub of the realist and humanistic critiques of positivism in human geography. To some, this has been interpreted as irrelevant because general laws do not preclude the existence of unique events (as Hartshorne, 1984, was at pains to stress with regard to his characterization of regional geography, so frequently misinterpreted by others). It was accepted that much of what was studied in human geography was particular to a place, an individual outcome (complex of phenomena making up a region, perhaps) not replicated elsewhere or elsewhen. But this individuality should not be equated with singularity. That individuality, or *uniqueness*, may be the consequence of a singular combination of circumstances, but only as a singular combination of general laws. We live in a complex world, and it is quite feasible that, with so many laws in operation, certain combinations are repeated only rarely, if at all. But this does not mean that the combination cannot be accounted for by general laws alone. It may be difficult to produce such an account, but if the basic argument is accepted then the general procedures of positivist science can be applied.

Such an argument would appear to be valid for physical geography, with few (if any) reservations. A very large number of physical, chemical and biological processes are operating at or close to the earth's surface, and they can be combined in an even larger number of ways. The processes are fixed and, as far as we can tell, subject to immutable laws; their interactions, too, are subject to similar laws (laws of wholes which may be more than sums of the laws of the parts). Eventually, therefore, it should be possible to explain all events as combinations of laws, and to explain how and why certain combinations come about.

But can such an argument be made for human geography and the social sciences, disciplines which focus on human activity? Are there similar general laws of behaviour, of reactions to stimuli, which also combine in many ways to produce unique outcomes? According to the humanistic critique the answer is no, the social scientist cannot treat people in the same way that physical scientists treat matter and machines; the analogy is invalid (see Harrison and Livingstone, 1982). Human beings have powers of thought and reason, which make them entirely different from inanimate matter – and those powers are many times greater than those of any other living species. Further, humans, individually and collectively, learn, are able to store their knowledge, and are able to recall it from independent memories (libraries, computer files etc.). They can draw on resources, which they interpret. Further, as the realist critique makes clear, the facility to learn means that the context for decision-taking is always changing. The future is not the past again, therefore, and to the extent that learning – what people remember and recall, and how they interpret those recollections – is unpredictable, so the future is constrained by, but not predictable from, the past.

How might the study of human geography be structured, given the recognition of separate places with different characteristics? The alternatives addressed in the literature are as follows.

1 To treat every event or outcome as *singular*, accountable only internally and with no reference to general laws. This was the implicit basis of much regional geography: the region was a singular place which could only be understood by studying it alone. As studies in systematic geography developed, this view was modified somewhat, because at least some of the parts – notably those in the physical landscape – could be portrayed as exemplars of general laws, and others – relating to human occupance (e.g. market towns) – were not peculiar to any particular region. But singularity was the combination of the parts, what gave the region its particular features, its regional character.

2 To treat events as *unique*, as accountable by a combination of general laws acting in concert. A particular place may be the only example of a certain combination, but there was nothing more to it than that, nothing that was singular to it. For human geographers, this meant identifying the laws, and the development of spatial science/locational analysis in the 1960s typified this approach; it was assumed, as made clear in the texts by Haggett (1965) and Morrill (1970), that there are general laws of spatial behaviour from which are produced general patterns of spatial organization. Places – or regions – were the outcomes of these laws, examples of the general models (von Thunen, Alonso, Hoover, Weber, Christaller, Palander, Burgess, Ponsard) of spatial arrangements discovered by geographers scouring the litera-

ture of the other social sciences. (The field developed, out of normative models and into behavioural ones, but the assumptions remained the same: Johnston, 1983a.) The result was a monocular view of the world.

For the realist approach to science, neither of these is valid.

The singularity approach is invalid because it entirely ignores mechanisms; the uniqueness approach is invalid because it assumes invariant mechanisms – laws. Positivist science is built on two types of law. The foundations are the *membership laws*, the allocation of items to general classes; hence Harvey's (1969, p. 326) statement that: 'Classification is, perhaps, *the* basic procedure by which we impose some sort of order.' Only when classification has taken place, when the membership laws have been established, can one move to the derivation of *functional laws*, the statements of relationships between classes of phenomena (manual workers vote for political parties that advocate socialism). As stressed here, the latter are invalid structurings of a changing world. Membership laws may be. To the extent that behaviour is learned, *in places*, then common characteristics are likely. But to classify people, and to treat them as exemplars of such classes, is to freeze a changing reality.

What, then, is the way forward? The answer proposed here is based on a realist interpretation of uniqueness. There are general laws, but these operate at the level of mechanism only, not in the empirical or the actual worlds. The actual world involves the interpretation of those laws, in the local context of contingently related conditions, and the outcome creates the empirical world. The outcomes of the laws are unique, not singular, because the individuals who put them into operation are, necessarily, acting in context. They are thinking, feeling human beings, and the only way to understand what they do is to understand them; to treat them in any other way is to deny their individuality and ignore their separate existence. They govern their own actions, in their personal environments; they think, argue, rationalize, feel emotions and assign meanings. But this is all done with the guidance of others: they can do those things because they have the intellectual equipment, but how they do them reflects their conceptual equipment, and this they obtain from others through environmental learning. They cannot be singular, but they are unique. They draw on resources provided by others (while at the same time providing resources for others) in order to interpret how they should enact the laws of the mechanisms, the imperatives of their mode of production.

Practical Science

The world may be in a state of constant flux, and therefore a positivist approach to the study of human geography irrelevant, but is this how people perceive it as they live in it? As already stressed several times, in order to live in a complex world people simplify it, creating a taken-for-granted world within which behaviour is habitual. Instead of engaging the complexity of milieux, people distance themselves from it, creating stereotyped – or purified, according to Sennett (1970) – images of reality within which action is organized. Just as positivist science is built on membership laws, on classifications of phenomena into types, so is everyday life – or practical science. These are not universal types, of course, since they are created by people in societies; they are shared meanings in places, as Ley (1974) makes clear in his discussion of residents' images of 'the world outside' their Philadelphia neighbourhood.

Can we not conclude that positivism is a valid approach to social science, therefore, because people act as positivists, creating images (membership and functional laws: 'that street is unsafe, and if I go into it there is a high probability that I will be mugged') and acting accordingly? But to equate practical science with positivist science is to assume that once created the images remain fixed. Ley (1974, p. 252) notes that 'some people are particularly susceptible to a hardening of the image', especially those he terms 'lower class individuals'. The image filters and codes information, it determines what people will reject and accept and, of the latter, how they will interpret it and react. And that filter is created through social interaction with others, usually others in the same locale, who provide the role models for socialization.

According to this view, life is segmented – people live within the confines of their group images, many of which are associated with a known, if ill-defined, turf. Their behaviour is habitual, and hence predictable. As a consequence, positivist analysis is possible, identifying the general laws of human behaviour in a particular setting.

But positivism freezes reality, and to treat images as such is to deny the possibility of change. Although people themselves may to some extent seek to do that, they cannot isolate themselves very far from external influences. Their milieux are always changing, in part as a reflection of what is going on outside their locales, in part because new people enter and others leave, and in part because they have not totally stopped learning; changes in the environment rarely allow them to. Thus they, individually and collectively, are always interacting with their milieux. The results of many of their interactions may be to allocate new information to the existing stereotypes. But this will not always be possible. Environmental change will create new situations that are not readily accommodated in the existing image, and so the image itself must be altered. Views of the world must be changed, and behaviour modified accordingly.

To the extent that they cannot accommodate environmental change within their present image, people act as hypothesis-testers. To know how to react to new information, they must test out its meanings. (To return to the earlier example of the unsafe street, people may be told that it is now safe – perhaps by police who have acted to remove the problem. And so they have a new hypothesis to test: that the probability of being mugged there is no greater than on other streets. They put it to the test, probably tentatively at first, and if they find the hypothesis valid, they respond accordingly by recategorizing the street.) Life, then, is one long process of environmental learning, of experimentation with cultural images and behaviour patterns. That experimentation very often involves 'field testing' of a hypothesis: as with so much learning, to be told something may produce little response, to be shown it may gain appreciation, but to experience it is more likely to lead to acceptance.

Only people with completely closed minds act as if they were positivist, therefore. Most act as hypothesis-testers, responding to changes in their environments by personal explorations which lead to revised, provisional rules for behaviour. Thus practical science cannot be equated with positivist science; it should, however, be equated with scientific rigour in hypothesis-testing.

QUANTIFICATION: MUCH MISUNDERSTOOD

To many geographers, science is equated not only with positivism but also with quantification; the three form an unbreakable trilogy. But just as the first equation has to be rejected, so do the other two: science need not involve quantification, and quantification does not necessarily imply positivism. Even were these misconceptions to be removed, however, there are still many other misunderstandings about quantification that form the basis of unjustifiable criticisms, and in consequence a wholesale rejection of quantification.

As an example of such criticisms, I use Andrew Sayer's (1984, p. 173) miscategorization: 'There are two types of statistics: descriptive; for example, measures of dispersion, and inferential; for example, regression analysis.' This is a totally false representation of the two types. Measures of dispersion – such as means and their associated standard deviations – are descriptive summaries of a data set, but they may be inferential descriptions, that is, estimates of the mean and standard deviation based on a sample. Similarly, regression analysis may be used in an inferential context, to suggest whether a relationship identified in a sample also holds in the population from which the sample was drawn, but it can also be used to describe without any inferential implications. (The regression of GNP per capita on population size for every country in the world describes the relationship between those two variables. Every country is studied, so there is no sample and no population to which the findings might be inferred.) Clearly, some clarification is needed.

Rather than categorize statistical procedures as descriptive or inferential it is better to use the typology of exploratory data analysis and confirmatory data analysis. The distinction between the two is clarified by Cox and Jones (1981, p. 135):

> In *exploratory data analysis*, attempts are made to identify the major features of a data set of interest and to generate ideas for further investigation, whereas in *confirmatory data analysis*, attention is focused on model specification, parameter estimation, hypothesis testing and firm decisions about data.

The two do not necessarily equate with descriptive and inferential. For example, a sample of the British electorate may have been questioned about their political opinions and activities, and the data set may be explored (variables cross-classified against each other etc.) in order to suggest relationships that should be the focus of further investigation. Those explorations are of a sample, and are describing what is present: they are not testing explicit hypotheses. But if the suggestions for further research are to be worthwhile, the analyst will probably ask whether the relationships identified are likely to be true of the entire population of electors; that is, inferential techniques are used.

Much quantitative work in geography is indeed exploratory data analysis, since the data available are not suitable for confirmatory data analysis, especially if the hypothesis being tested, or the model being calibrated, is assumed to have wider application. In part this is because of the nature of the sample being used. Two main types can be identified:

1 The explicit sample involves the careful definition of the relevant population, from which a sample is then drawn according to the accepted rules; and

2   The implicit sample involves the study of the population in a particular place, from which general conclusions regarding a wider population may be drawn.

The first type is the exemplar for inferential statistics (though note that a hypothesis does *not* have to be tested on a sample); it is relatively rare in human geography, however, certainly so in other than survey analyses. The second type is the most common. The researcher may have a hypothesis relating, say, to the settlement pattern in a rural area. This is tested in a study area. What if the hypothesis has been found valid? It has been tested on a single sample and, as all statistics texts show, the error terms associated with samples of one are very large; it tells us nothing in general, although perhaps plenty about the particular case. (Note that this is a particular problem with regard to the positivist practice of testing a hypothesis by seeking to validate it, which leads to the problem of inductivism; however many positive tests you get, you can never be sure that the next will not be negative. Thus some argue – e.g. Marshall, 1985 – that the methodology of critical rationalism, developed by Sir Karl Popper, is much to be preferred. Instead of seeking to validate a hypothesis you seek to refute it, with a single refutation being sufficient to cause the hypothesis to be rejected in its present form. This avoids the problem of inductivism. But what is a refutation? Does an R^2 of 0.3 refute a hypothesis?)

Much geographical work is concerned with implicit samples only, therefore, which can be used to refute but not to validate hypotheses. Does this mean that inferential statistics are largely irrelevant to human geography, therefore? The explicit sample is not being used, and the population may be unknown: generalization from sample to population is then impossible. But this is not the only use for inferential statistics. Inferential procedures can be used without any transfer of findings from sample to population. Take a data set relating to observations of land value and distance from a fixed point in a city. Exploration of these data wants to know if there is a relationship between the two, and how strong it is. There is a finite number of permutations of the two vectors of data – random pairings of distance and land values – each of which will yield a correlation coefficient between the two variables. To assess the observed value, it can be set in the context of the distribution of all possible values. If it is a relatively rare event, one might conclude that it was unlikely to have occurred as the result of a random pairing of the data points. A statistical significance test – usually associated with inferential procedures – could be used to assess the rarity value of the observed correlation (remember a significance level of 0.05 is an entirely arbitrary criterion, a guide, but no more, to rarity value), and so indicate whether what has been observed is worthy of further investigation; inferential statistics have been used in an exploratory context, either in the direction of future research or in the evaluation of a description. (We should note Stan Openshaw's important work on the modifiable areal unit problem, however, which shows not only that our findings are probably a function of the areal units that we use but also that manipulation of those units could produce the results we want: Openshaw and Taylor, 1981.)

Inferential statistics have a proper role to play in story-telling, therefore. Furthermore, it has not to be assumed that descriptive story-telling is necessarily to be based on 'simple' statistics only. Sayer assumes that descriptive statistics are simple (of the type taught in introductory courses) but that the more sophisticated (second-year courses) are necessarily inferential. In fact, most of the inferential work reported relates to the use of the simpler statistics: how often are statistical significance levels reported in discussions of factor analyses as compared with presentation of chi-squared analyses of 2 x 2 contingency tables? And relatively sophisticated procedures can be used in exploratory descriptions, as in the application of entropy-maximizing procedures to derive maximum likelihood estimates of unknown values in cross-classifications (for example, the work on estimating vote by occupation and housing tenure in each English constituency in 1983: Johnston, 1985a).

Quantitative procedures are too frequently used wrongly, to provide what might be termed a pseudoscientific gloss. But this is a problem of the usage, not of the procedures themselves. Criticisms of such usage are clearly right. But this should not form the basis for outright condemnation of quantification in human geography. Story-telling involves using the available texts to weave a coherent story about an empirical topic, in the context of an overall theoretical perspective. Those texts may well be data sets and relevant procedures, descriptive or inferential, exploratory or confirmatory, should be used to the maximum to enrich the story-telling, especially in those contexts where, as Pickles expressed it (see p. 72), the outsider can provide a different (better? certainly broader) interpretation than the insider. As Taylor (1981b, p. 265) expresses it in his defence of factor analysis, and as he demonstrates superbly in his materialist interpretation of US Presidential elections (Archer and Taylor, 1981): 'The technique is merely a measurement tool to define a concept or concepts within a clearly specified social model.'

There are, of course, other criticisms of quantification, many of which rest on the claim that human characteristics and variability cannot be reduced to numbers or treated as 'things' (Giddens's first proposition;

p. 83). Certainly there are many, probably most, aspects of the worlds of experience and events in which this is so and the application of quantitative procedures wrong. But the existence of some wrong usages should not lead to a condemnation of all uses. There are aspects of life in which numbers are the outcome, many of them involving classifications. To return to Peter Taylor's example, elections necessarily classify us (in the USA the vast majority of adults fall in one of the following four categories: Republican voter; Democrat voter; registered voter who abstained; unregistered), and, as recent work has made clear (e.g. Wrigley, 1985), much can be learned from the quantitative analysis of classifications (although some of the models tested are based on positivistic theory – of utility, for example).

IN SUMMARY

Positivism as a philosophy of social science cannot be accommodated within the realist approach adopted here, even in some of its 'milder' forms (for example, not making the scientism and value-freedom claims noted by Keat – p. 84 – and not assuming that all aspects of human life can be categorized according to membership laws). It assumes, and seeks to identify, laws, which is a false interpretation of how human agents learn to live in a constantly evolving dialectic between worlds of experiences and worlds of mechanisms. Positivism freezes the world into invariant processes of change: realism sees it for what it is, a world of continuous change within a context of mechanisms.

In rejecting positivism, some human geographers have rejected both rigorous hypothesis-testing and the application of quantitative methods, procedures which entered the mainstream of geography contemporaneously with the adoption of the search for laws. That coincidence was unfortunate, for it has led to the rejection of procedures and methods that are necessary to, but not solely associated with, positivism. Both are certainly valid within a realist approach, as argued here. Rigorous scientific practice through the testing of hypotheses is characteristic of daily life in the world of events (in the abstract as well as the actual, as with business executives simulating the likely consequences of their decisions, perhaps through large computer models), and is an important means of assembling material from texts into a coherent portrayal of the world. Quantitative procedures, too, are important to some aspects of story-telling, and have too frequently been misrepresented. To reject positivism, an end, must not be equated with rejecting certain means that are not the sole preserve of that approach.

6

Applied and Applicable

The commitment to social relevance in geography has a long and distinguished history, which can only be dissected by patient reconstruction of past intellectual and social endeavours.

D.R. Stoddart, 1981, p. 5

There have been many calls from human geographers in recent years for their discipline to become more applied, or 'policy-relevant', in the orientation of its research and in the goals set for its teaching activities. There have also been plaintive cries that geographical expertise is insufficiently appreciated, in part because outsiders have an obsolete view of the discipline and the expertise at its command and in part because geographers are considered not to have sold themselves very well in the past. None of this is new, as the quotation above indicates; indeed, Michael Wise (1984) has referred to applied geography in the UK as 'surely one of the strengths of the subject'. But there is no doubt that the felt need has increased, a need reflected, for example, in the convening of annual conferences on applied geography in the USA and the foundation of a journal, *Applied Geography*, in Britain in 1981.

Why, then, is there such pressure at the present time? The root cause is the economic recession (some would say crisis) that has been felt throughout the world-economy in the last two decades, and which has affected Britain particularly badly. From this, two sets of pressures have emerged, one internally generated and the other externally imposed.

The internally generated pressures reflect the responses of human geographers as citizens to the economic and social problems that they perceive, and which the development of welfare geography (D.M. Smith, 1977) has impressed upon them. Not unnaturally, some human geographers have been moved, perhaps by perceptions of self-interest but more likely as a moral response to the evidence, to orient their work in some way towards the solution of those problems. In so doing, by acting in what they perceived as a relevant and desirable way, they may have been concerned that what they did was so scathingly treated by some of their fellows. David Harvey (1973), for example, told them that their proper programme of work

> does not entail yet another empirical investigation ... In fact, mapping even more evidence of man's patent inhumanity to man is counter-revolutionary ... There is already enough information. (p. 144)

He called on them, instead, to develop a deeper awareness of the processes that produced such inhumanity, and apply that not in support of the proto-fascist, corporate state (1974, p. 23) but rather, as he expressed it a decade later (1984), to:

> Create an applied people's geography, unbeholden to narrow or powerful special interests, but broadly democratic in its conception ... [as part of] a political project that sees the transition from capitalism to socialism in historico-geographical terms. (pp. 9–10)

Nor were they pleased when Zelinsky (1975) concluded in his Presidential Address to the Association of American Geographers:

> how woefully deficient we are in terms of practitioners, in terms of both quantity and quality, how we are still lacking in relevant techniques, but most of all that we are totally at sea in terms of ideology, theory and proper institutional arrangements. (p. 529)

But these were isolated, though powerful, voices. Others were urging them forwards, though cautioning care in not offering more than could be delivered.

The external pressures resulted from government interpretations of the economic recession and theories of how to emerge from it. In both the United Kingdom and the United States, the solution to the problems advanced by governments since the late 1970s has been based on firm beliefs in free-market capitalism. The task of the state is to promote accumulation by controlling inflation and the money supply, encouraging enterprise and investment, and legitimizing a relative decline in the provision of public services and the welfare state. Further, governments have been convinced of the need for massive investment in research into and development of new technology. To achieve these goals, the higher education budget has been cut and there has been a redirection of the available resources towards science and technology. Universities and other institutions have been encouraged to emphasize the *training* of students (the provision of 'saleable' skills) over the promotion of general *education*, and to seek a greater proportion of their income as research contracts and commissions from bodies other than the state.

Members of all disciplines have been affected by these trends, but some – like human geographers – have felt more vulnerable than others because of their generally poor performance, as they saw it, in terms of fund-raising and producing saleable research and graduates. They have felt the need not only to reorient their work (emphasizing, for example, technological expertise in: data acquisition – remote sensing; collation – geographical information systems; analysis – spatial statistics; and presentation – automated cartography) but also to polish their public image and present themselves more forcefully to the public, to business, and to educational policy-makers.

Applied geography is being presented today as both necessary to the future of the discipline and as a contributor to the solution of economic and social problems. Geography is being placed in the service of society. But is it not a rather narrow definition of service that is being promoted? The present chapter seeks a broad evaluation of applied human geography.

APPLIED GOALS

The term 'applied geography' implies utility, that work is being conducted in a context that involves its immediate use by and for society, hence the emphasis on training rather than education. But what is useful, and for whom? Before evaluating applied geography it is necessary to take a wider look at how science might be applied. For this, science is divided again into three broad types: empiricist/positivist, humanistic and realist.

Empiricist/Positivist Science and Technical Control

Empiricist science operates in the world of experience only, and most of it is based on an assumed neutral, outside observer. Positivist science uses the data obtained in this way, and structures it into general laws. The goal is *explanation*, interpreted in this context not just as 'providing an account for' but rather as 'accounting for individual events as examples of general classes of such events'. The laws that provide the explanation identify the antecedent causes of the observed events, thereby equating explanation with prediction: one can account for a phenomenon because one can say in what circumstances it will occur.

Use of knowledge produced in this way is associated with an ideology of *technical control.* The laws can be used to predict (estimate or forecast) events, so that it is possible to evaluate the consequences of certain actions and to work backwards from a particular goal to a set of processes that will lead to it. With such laws, therefore, it is possible (a) to avoid certain types of event by ensuring that their preconditions do not occur, and (b) to ensure that other events happen, by engineering their preconditions. When such an approach involves applied social science then the product can be termed *social engineering*, the manipulation of society towards certain ends and away from others.

Such application of knowledge is inherently conservative, since it takes the empirical superstructure of society as given, identifies its empirical laws, and uses them to reproduce society. Change is excluded, apart from that change which conforms to the laws. Policies of technical control based on empiricist/positivist science are thus based on an acceptance of the current structure of society.

Humanistic Science and Mutual Understanding

The goal of the humanistic sciences is to achieve an understanding of events, the thoughts behind the actions that produced the world of experience. The only explanation that it seeks is the account provided (directly or indirectly – through texts) by the actors themselves. There is no future-oriented goal, only appreciation of what created the present – when it was created.

Such sciences have a clearly defined applied goal, however: to increase both *self-awareness* and *mutual awareness*. Self-awareness is brought about by assisting people to reflect on their situation; mutual awareness by promoting appreciation of the situations of others. Thus hermeneutics are involved in the transmission of information.

In such applied work, the role of the researcher is not that of a technician, one who promotes a certain solution, but rather that of a *provocateur*, one who promotes thought and reflection. The goal is non-prescriptive, because people are not being directed towards a particular end; self-control remains unimpaired, but its information base is influenced by disinterested outsiders. A major problem in this is in communication. Language itself is a barrier, as Gunnar Olsson (1982) has (infuriatingly!) demonstrated. But the expression of ideas is even more of a barrier. Different perceptions of the same thing can lead to much unproductive debate, as frequently illustrated in international and industrial relations. As a consequence, 'solutions' are often only achieved when a more powerful group imposes its will on others. 'Resolutions' require mutual understanding, through processes of mediation whereby each party learns more fully about the others and itself, whether through the offices of shuttle diplomats, industrial conciliators or marriage guidance counsellors – people who assist others to make decisions but do not make them for them.

Realist Science and Emancipation

The mutual understanding promoted by humanistic science may be mutual appreciation of mutual ignorance (though the 'fact' that it is ignorance is, of course, not appreciated). People have their perceptions of the world and how it works, and these can usefully be transmitted to others. But if those appreciations are at best cloudy, at worst false, then becoming aware of each other's ignorance may be interesting and perhaps beneficial, but ultimately unproductive. Much more vital, according to the realist conception of science, is removing the falseness, dispersing the clouds, and promoting real understanding.

As outlined in chapter 4, real understanding can only be achieved by setting the empirical and actual worlds in the context of their containing structures, the world of the mechanisms of the economic base and the societal superstructure. Distorted views of those structures (probably ideologically created and promoted) must be removed and replaced by real understanding, by explanations of events that can account for their occurrence as individual operations of the mechanisms, but not as examples of general empirical laws. Such understanding must be achieved by those 'working with the tools of the academic trade' (Harvey, 1973, p. 167) but then transmitted to members of society, who should be involved in development of the understanding (hence the people's geography).

This is the goal of *critical theory*. It, too, uses hermeneutic procedures to increase awareness, but of the real as well as of the actual and the empirical, and of the links between the three. As Bernstein (1985, p. 35) expresses it: 'We cannot understand the character of the life-world unless we understand the social systems that shape it, and we cannot understand social systems unless we see how they arise out of the activities of social agents.' The other two ideologies obscure rather than reveal: that of empiricist/positivist science suggests that societies can control their futures and produce 'better worlds' through empirical manipulations; that of humanistic science suggests that people, as individuals, can create their own futures. Realist work is *emancipatory*, releasing people from these false ideologies by identifying the mechanisms, indicating the constraints that these impose, and showing that only a replacement of the mechanisms can remove those constraints. The goal of emancipation is social change – neither the cosmetic social change associated with technical control, which reproduces the mechanisms, nor the romantic social change associated with the voluntarist ideal of awareness, which ignores the mechanisms' existence. It aims for fundamental social change, through the replacement of the mechanisms.

There is, then, not one but at least three types of applied science. All of them are relevant to the practice of geography, as the next sections will illustrate.

APPLIED GEOGRAPHY AND TECHNICAL CONTROL

Much explanation in the positivist sciences is expressed as causal laws of the 'if x then y' variety. In most cases, these have been derived in laboratory situations, with contingently related conditions held constant. In the 'real world' many such laws are operating, separately and in combination, so that explanation of empirical events necessarily requires complex models of systems of interacting laws. To apply the results of positivist science, therefore, it is necessary to go beyond the task of abstracting one or a few links from such systems and conducting laboratory experiments. The systems as wholes must be modelled, and their total operations subjected to experimental work. Only then will it be possible to answer questions such as:

> If we change the value for this element in the system, and the strength of that link, what will be the effects (direct and indirect) on all of the other elements and links in the system?

and

> If we want to reach a certain goal – a predefined value for a given element in the system – what other elements and links will have to be altered, and what will be the other consequences of those alterations?

The first sort of question is asked by those evaluating the impact of an increase in the release of radioactive effluent from power stations, for example, whereas the second is asked by those considering how such effluent might be decreased.

Positivist natural science is organized to tackle such questions through its law-seeking procedures. It models the systems, identifies the basic inter-relationships within them by experimenting on the parts in artificial conditions, and uses the laws discovered to build up models (usually sets of equations) describing the systems as wholes. There are many difficulties en route, and the separate claims of reductionism (that a system is merely the sum of its parts) and of holism (that there are laws of the whole systems as well as of the parts) are the cause of much debate (see Simmons and Cox, 1985). But there is a strong belief that full explanation is possible, if hard to achieve. (This is at the empirical and actual levels only. A realist theory of science, similar to the realist theory of social science advocated here, points out that no amount of empirical work can elucidate the mechanisms themselves and why they are present; you can identify regularities that you associate with the gravity model, but you can never observe gravity itself or know why it operates: see Johnston, 1986b.)

Physical geographers have adopted this perspective in large numbers in recent years, with a shift away from the description of forms and from essays on the long-term evolution of landscapes (built on the Darwinian roots of geology) and towards the elucidation of contemporary environmental processes. As such, it has been able to develop an applied orientation, providing evaluations of the likely impacts of interruptions to, or changes in, environmental processes. These are based on careful experimental work, in laboratory and controlled field situations, the results of which are fed into larger models of environmental systems.

Development of explanations as particular laws and of calibrations of system models has been slow, because of the complexity of the subject matter. The number of elements and links operating in the physical environment is so large, and the nature of the inter-relationships so complex (involving not only the quantitative change associated with linear models but also the qualitative change represented by models of discontinuities, such as catastrophe theory and bifurcation), that it is very difficult to identify a rational abstraction, a part of the system that can be treated as if it were closed, that is, with few salient external links. So many things have to be held constant in the abstraction that the utility of the results of laboratory experiments is frequently dubious, whereas the results of field experiments are often inconclusive because contingently related conditions could be neither held constant nor eliminated. Climatology illustrates this well. Laboratory experiments are mainly irrelevant, and it has been necessary to develop large models of a complexly interacting, global system. Field measurements designed to contribute to the calibration of those models cannot take account of contingently related conditions, and so it is difficult to obtain reliable parameters for the many links in the systems. Simulations can provide general understanding, but the number of possible combinations of contingently related conditions is so large (and crucial to the weather produced, and hence to prediction) that full understanding of particular events is still a long way off. The task is feasible, but massive.

In physical geography, positivist scientific explanation leading to prediction and thence to environmental manipulation and control is the goal; it is possible, if unlikely to be fully successful. Is such a goal feasible for human geography too? Is it possible to identify the laws of human behaviour, integrate them into models of societal systems, and so develop the practice of spatial social and economic engineering?

The answer widely held within contemporary society to the last question is yes. There is a general belief that such social engineering is possible. We live in what can be termed a *pseudopositivist world*, one in which the general tenets of positivist science are applied to human actions and societal operations; this is a belief in what Keat (p. 84) has termed scientific politics. Thus governments and their advisers have models of the operation of economic systems, for example, which contain many elements and links and are represented by large sets of simultaneous equations (as illustrated in Bennett and Chorley, 1978). They then ask questions of the models like the second one above (p. 105), and from the answers deduce what their economic policies should be: which elements in the system (the money supply, the rate of interest) or which links (types of investment, saving or spending) should be manipulated to produce a desired goal (low inflation, low unemployment, increase in GNP etc.). And this belief in scientific politics by those who practise it is conveyed to society as a whole through the state ideology (p. 35), creating a widespread belief not only that the government should do something about the issues facing society but also that it can.

Such activities are normally encapsulated in the term planning, which for human geographers implies planning of spatial organization and environmental use. There is a long record of geographical contributions to such activity, largely involved with collection of the data on which planning could be based. With the development of a positivist human geography and the representation of spatial organization in systems terms (as in the works of Mike Batty, 1976, and Alan Wilson, 1974, for example), geographers have become more fully involved in providing the tools for planning as well as its information base. Brian Berry (1973) has identified four types of planning activity in this context.

1   *Ameliorative problem-solving – or planning for the present.* This involves the identification of a problem which requires immediate

action – serious overcrowding of housing in an area, say, or a major traffic bottleneck – and the relevant local parts of the system are manipulated to provide a ready solution, without much attention to the likely feedforward and feedback consequences. It is likely that the problem will reappear, perhaps in a slightly different form, perhaps elsewhere, soon, but meanwhile the immediate issue (and its possible effect on government legitimacy) has been removed.

2   *Allocative trend-modifying – planning toward the future.* In this activity, major trends within a system are identified, and resources are then allocated so as to accentuate their more desirable aspects; the system is steered in the general direction it was heading for in any case. Thus likely problems are identified (a demand for new routes, for example) and avoided.

3   *Exploitative opportunity-seeking: planning with the future.* This is similar to the previous type, except that little attention is paid to the possible deleterious consequences of the identified trends. Whereas trend-modifying suggests the articulation of a clear goal, in the context of current events, opportunity-seeking aims merely to reap whatever benefits are available, and is inclined to let the future take care of itself.

4   *Normative goal-oriented planning for the future.* Unlike the other three, this does not start with the present and project forward: it starts with the future and works back, identifying what needs to be done in order to reach a predefined goal.

All of these activities involve creating the future. But there is also the need to monitor the future, as it happens. Geographical involvement includes the following activities:

1   *Analysing intended policy impacts,* which involves asking whether a policy has worked, in terms of the goal set. Much of this is relatively straightforward (the policy intended a certain outcome; did it eventuate?), although there are problems of knowing what would have happened if the policy had not been enacted – what is the datum against which the policy should be evaluated?

2   *Analysing spatial variations in policy implementation and impact,* which is an extension of the first type, looking more specifically at distributional aspects. Many policies involve the public provision of

goods and services at fixed places; others involve their provision by areas (perhaps by local agencies to which responsibility is devolved). And many policies are aimed only indirectly at the individuals who are supposed to benefit; they are the targets but the implementation is only to the area in which they live (the aspirin approach). To what extent, then, are they successful in reaching the intended beneficiaries, or do some people, in some places, get an advantage over others that was not intended.

3   *The analysis of unintended consequences* which come about because all of the impacts of a change in the system have not been evaluated before policy implementation (as in ameliorative problem-solving). This requires a properly specified model of the system involved.

4   *The prediction of future policy impacts*, which involves prior evaluations similar to the first three (except that they are posterior evaluations).

The existence and promotion of all of these activities implies that technical control is possible in this context, that one can have an applied positivist human geography involved in the creation of the future. There are several arguments against this, however. The first is not a fundamental one, but only operational. It is similar to the case presented above regarding the difficulty of modelling climatic systems and hence producing accurate weather forecasts, but it is more defeatist. In human geography, everything is linked to everything else and everywhere is linked to everywhere else, so that abstraction of a part is at best very partial and at worst an entirely counter-productive chaotic conception. To study just one piece out of context is to misrepresent reality grossly, and serve no purpose, because the external links are so important, and yet unconsidered, that any forecasting and prediction is (logically) impossible. To accept this argument is to agree that much planning cannot be undertaken in the positivist context. But are all abstractions from socio-economic systems necessarily chaotic conceptions? To study one town's economy without reference to movements beyond its borders certainly is; to study its traffic problems may not be.

A second, more fundamental, argument is that planning through the use of statistical laws and mathematical models uses a machine metaphor which is irrelevant to the analysis of human societies. Humans are thinking beings, not automatons, and to treat them as programmed responders to stimuli denies their humanity. Furthermore, and more practically, to model them in that way is counter-productive because future behaviour cannot (or should not) be constrained by present behaviour. People can and do change, and social engineering should neither assume that they do not (with consequences for the validity of the models) nor ensure that they will not (either physically or ideologically). And because they do seem to act habitually in a taken-for-granted world should not cloud the fact (p. 94) that they are also rigorous hypothesis-testers.

Following from this argument is a third, that indeed people are changing all the time. Every action contains within it elements of change, and the society is not the same after that action as it was before. And, crucially, the nature of that change is not predictable, unless a great deal is known about the contingent conditions, now and in the future – as the steel company closure example showed (p. 78). Thus modelling, even dynamic modelling, of society is irrelevant because it refers to an historically (in time and place) specific instant. One can learn from the past, but no more.

Nevertheless, there is much planning and planning-related activity taking place, and human geographers are being encouraged to get fully involved in it. The pseudo-positivist world is in operation. Furthermore, many argue, even though one accepts the force of the above criticisms it is necessary to act in this way, because the consequences of not doing so would be a much worse society than we have and are getting. We need to be prepared, whether for a massive flood of migrants (who might not eventually come) or a major increase in road traffic, for example. The information on which we base those preparations is imperfect, but it is not useless (except where it is self-defeating, as in traffic forecasts and the consequent road provision). And the work done in recent years does indicate habitual behaviour, especially in certain contexts. Some people will undoubtedly 'deviate' from model predictions, but then the models were only supposed to be valid at an aggregate level.

Such responses may well be bolstered by reference to the sorts of work that human geographers are involved in. Much of it is concerned with the local scale of experience only, with aspects of daily life which are important to individual short-term well-being but trivial in terms of the trajectory of the mode of production. Access to shopping centres, the provision of country parks, and the details of a bus route are of local, transitory importance only, but nevertheless attention to them is not entirely irrelevant, since if planning can improve access and provision it will help to maintain social consensus and thus aid the state in its legitimator role. (There is an interesting paradox here. In the mid-1970s the British government decided it would be good if people were provided with information on grocery prices, so that they could make more rational choices of where to shop. This was duly done, and weekly surveys of the price of a standard basket of goods in various shops were conducted and published. They showed, not surprisingly, that some shops were cheaper than others for that basket. But further analyses that Alan Hay and I did – Johnston and Hay, 1979 – showed that the rank order of the shops on cheapness varied substantially according to which sample of goods from that basket was being bought, and our modelling – Hay and Johnston, 1979, 1980 – showed that such an uncertain environment was clearly in the shopkeepers' interests. So unless they used the information provided in a very sophisticated way – or bought only the basket of goods specified – shoppers could lose, not benefit. The information is no longer provided.)

Thus many human geographers who work on problems relating to everyday life can claim that they have a valid role as positivist-technicians because it is better for society to operate on the *as if* assumption that the pseudopositivist approach is valid than to be both unprepared and in danger of social dissent if everyday difficulties are not attended to. This is a conservative position to take. It accepts the basic structure of society, and seeks to manipulate certain aspects of its superstructure only; it accepts the need for a strong state to implement planning; and it realizes that its major contributions are likely to be in ameliorative problem-solving only. It is involved in patching-up the future, rather than creating it.

And yet, it could be argued, such an approach to date has had massive benefits, for life expectancy levels have never been higher and the standard of living/quality of life never better. There are still major inequalities in the world, but the success of capitalism is slowly raising everybody's life chances. We would do well to give the market as free a rein as possible, so that this can continue. Against that is the argument that relative inequalities are not being reduced very much, that immiseration continues on a very large scale (currently in much of Africa), that the environment is continually being degraded, and that the end of optimism is nigh. To treat the symptoms rather than the cause of problems not only simply redistributes the problems it also aids and abets the interests of those within the global economy who would suffer from the effects of fundamental changes which tackled the problems at source. We can suggest how central government money should be allocated to local governments in order to meet felt needs, but we are not able to tackle those needs themselves and eradicate them.

Choice of the role of applied human geographer as technician is necessarily an ideological one, therefore. If the scientists claim that they are neutral and that the use of their work is a decision of others, that the work is applicable rather than applied, then they are bolstering the status quo. By being positivist scientists they are creating knowledge which can be used. The dilemmas that this poses are returned to at the end of the chapter.

APPLIED GEOGRAPHY AND MUTUAL UNDERSTANDING

The single world-economy is divided up into a large number of sovereign states, within (and overlapping) which there are in many cases two or more separate social formations; state and place are rarely spatially coincident. The majority of the residents of those places and states have no direct knowledge of any of the others – and the vast majority of the rest know only two or three. And yet economically their lives are inextricably interlinked in the global economy. Indeed, even within states, many people have little contact with the great majority of their fellow citizens, and exactly the same can be said of the residents of cities and towns. Thus the economic system operates in a situation of mutual ignorance.

In general terms, such ignorance is only to be expected, given the distances and the numbers of people involved. But how do people react to their ignorance?

As already noted, in all aspects of life people react to complexity by creating simplified models of it: ideal types. This is as true with regard to reactions to people as it is to other stimuli. Types are defined on easily used criteria (race, gender, age, language etc.) and individuals are allocated to the relevant category. Behaviour patterns are a consequence of the typification, because, as Sennett makes clear in *The Uses of Disorder* (1970), the types are not objective categories, as you would expect to find in positivist science, but the repositories of meanings. Those meanings usually involve both positive and negative connotations, with each type being contrasted with the individuals' views of their own type, their self-image. Individuals are then treated not as individuals in their own rights but as representatives of the stereotype. (This is especially so in impersonal treatments. Some people never encounter representatives of some of their stereotypes directly. Others do, occasionally. They may find, if they are prepared to get to know the individuals, that they do not fit the stereotype. But the stereotype is not altered; the individuals are treated as deviants.)

The development of stereotypes is part of the socialization process that is most active in the early years of life and is strongly influenced by attitudes at home, in the local area, and at school as well – increasingly – as by the media. There are local stereotypes therefore (often expressed through humour: to Britons the Irish are the focus of much humour; within Ireland, their equivalent of 'Irish jokes' refer to the natives of Kerry). But perhaps more importantly there are national ones, developed through the state ideology (of country versus country in capitalist competition) and fostered (often implicitly, but sometimes explicitly) through state education systems and state-sponsored media. Not all of those stereotypes refer specifically to places, for much of the internal operation of a society is facilitated by stereotypes related to, for example, class, gender and age. Even with these, however, there is spatial distancing: the class-based urban residential patterns, for example, and the gender-based allocation of space at several scales (within homes – kitchen *v.* study; workplaces – office *v.* workshop; cities – CBD *v.* suburbs, etc.).

The creation of stereotypes, and their relationships to the use of space, is part of the taken-for-granted worlds in which people live as pseudopositivists. The results of this can be disturbing, because they are part of what Sennett calls 'purified identities' which lead to immature behaviour that can trigger conflict, especially since those purified identities refer to communities – usually the residents of places, including states. The immature behaviour involves an unwillingness to face the unknown; rather than confront differences and accommodate to them, we retreat behind barriers. National borders are examples of those barriers, as are the 'frontiers' between the defined, yet unmarked, territories of various groups at smaller scales, such as neighbourhoods. Such barriers must be porous, however, to allow the ebb and flow of economic life, thereby producing contact. For much of the time, contact will be kept to a minimum and an uneasy accommodation maintained. But when difficulties arise, people may be unable to cope. As Sennett expresses it:

> Individuals in the community have achieved a coherent sense of themselves precisely by avoiding painful experiences, disordered confrontations and experiments ... Having so little tolerance for disorder in their own lives, and having shut themselves off so that they have little experience of disorder as well, the eruption of social tension becomes a situation in which the ultimate methods of aggression, violent force and reprisal, seem to become not only justified, but life-preserving. (pp. 44–5)

Stereotyping and distancing contribute to neighbourhood conflict at one scale, to nationalist conflict at another, and to inter-state wars at a third.

Sennett's development of this theme was undertaken in the context of the US race riots of the 1960s, which he linked to urban residential segregation. The ignorance and tension provide the context within which serious conflict could be triggered by any range of circumstances, however minor, as was the case in British cities in 1981. His proposed solution was to break down the purified identities and the stereotypes by breaking down the ghettos, mixing people of different backgrounds together and forcing them to reach accommodations: 'the experience of living with diverse groups has its power. The enemies lose their clear image, because every day one sees so many people who are alien but who are not all alien in the same way.' (p. 156)

Whether Sennett's proposal is workable is very dubious, but in any case it is not applicable at larger scales. And yet the tension between places at the state scale is of even greater potential importance. Today, much more than ever before, an eruption could lead not just to war, not just to the destruction of societies, but to the destruction of the whole of humankind, and indeed the earth itself as a sympathetic environment for human life. We have the capacity to destroy ourselves many times over at the mere push of a button, and that event could be triggered by a chance encounter, bred in a climate of ignorance, fostered by immature behaviour, and stimulated by nationalist ideology.

How, then, do we reduce the tension, promote accommodation, create maturity, and defeat xenophobic ideologies? At the local and even national scales, people can learn to appreciate diversity, to accept differences and coexist in harmony, through contact. But at the larger scales, this is impossible. Accommodation must be promoted by education, by creating positive attitudes to peoples and cultures, not negative ones. For this task, human geography has a clear role, as the discipline best able to display and describe the diversity that is the cultural geography of the world. A human geography should be humanitarian.

Implementation of applied geography as the promotion of mutual awareness is very largely an educational issue, a topic taken up in chapter 8. But you cannot apply ignorance. Human geographers – British human geographers in particular in my context – are very myopic in their research activity at the present time; as I have expressed it elsewhere (Johnston, 1985b), they have disengaged themselves from overseas research. (In part this reflects diminished opportunities for such research, a consequence of state spending policies and parochial ideology; see Thrift, 1985.) There is a clear need for much more research into other places, not in the positivist mould of seeking to account for what happens there in terms of general laws (most of which will reflect an imposition of outsiders' views of what is 'right') but in the humanistic sense of seeking to appreciate how societies are structured and to transmit that appreciation to others. The world is our oyster, but unless we promote its proper study it may soon be nobody's.

As an aside at this stage, there is a major applied task for physical geographers in a similar context, promoting awareness of environmental constraints. The positivist approach to the environment implies technical control; it is associated with what O'Riordan (1977) calls a technocentric ideology, built on so-called professional expertise, presenting itself as rational, objective and efficient, but in effect arrogant and elitist, and believing in the positive value of objective science and intervention. Against this he sets the ecocentric ideology, based on humility in the face of environmental processes, which recognizes that humans cannot control the environment, only destroy it. The sorts of environmental education linked to these ideologies are very different. The one promotes a vision of control, the other of accommodation; the former dominates, but its optimism is proving unfounded (Mercer, 1985a).

The foundation for such an educational enterprise will be a 'new regional geography' (Johnston, 1984a, 1986a). This will not be the descriptive trivia and catalogues of earlier regional geographies, but an engagement with the cultural diversity of the world in the context of the realist approach in which places matter. By promoting awareness, it may also stimulate emancipation.

APPLIED GEOGRAPHY AND EMANCIPATION

Mutual understanding at all levels – individual to international – may promote relative peace and harmony, counter tension and conflict, and reduce military posturing. Thus an applied geography which advances mutual understanding is to be welcomed and sustained. But the ignorance that it dispels is at the empirical and the actual levels only; what of the real level?

The cultural geography of the world reflects the development of separate societies in relative isolation and their subsequent contact with others, with many hybrids as a consequence. They have developed in the context of not only the environmental and societal interpretations but also the containing structures that they have put in place and subsequently sustained – their modes of production. Full understanding of societies, then, should involve seeing them in their historical context, as interpretations of the imperatives of one or more mode of production.

The goal of emancipation involves revealing to individuals, and thus to societies as a whole, the forces that underpin how their own and other societies operate. Those forces are not directly apprehendable; you cannot observe the accumulation imperative any more than you can see the law of gravity, all you record are events that are consistent with both. Thus appreciation of the structure of society involves: (1) the construction of adequate theories of the world of mechanisms; (2) advocacy of a non-determinist, non-voluntarist approach to understanding and explanation; and (3) development of the skills of story-telling, integrating their study of the real, the actual and the empirical in a coherent narrative. In this way, research results can be used to reveal not only the workings of our own social formation but also those of others, portraying them not simply as interesting 'deviants' but as separate interpretations of the same basic forces.

In such an approach the role of empirical and actual investigations is not just to provide illustrations of the distinctness of separate social formations but also to bolster the theoretical appreciation of the mechanisms in the real. For example, the Sheffield steel closure case illustrates the map of choices available to the state in such situations and provides a framework for studying which choice was made. Comparative analysis with other cases can be used to illustrate: (a) how different choices result from different contexts; (b) how whatever choice is taken, the longer-term goal of the state is to promote and legitimate capital accumulation, so that the differences between the cases are of means, not ends; and (c) how the contingently related conditions are therefore crucial for the form of the choice but not for its rationale. There is, then, a great need for comparative work, in the style of the 'new regional geography' advocated above, with the goal being not just to show that things are different in different places (or at different times in the same place) but rather to show that the world is a mosaic comprising spatially separated (though linked) groups of people working out solutions to similar problems. Every capitalist state, it is argued, needs to promote and legitimate the accumulation imperative and to maintain a social consensus accordingly. Applied geography as emancipation seeks not only to promote awareness of how this is being done in different places, but also why.

A major problem facing the emancipatory task is that the mechanisms at the real level are difficult to comprehend by many not only because they are not empirically apprehendable but also because ruling ideologies interpret them in particular ways which are consistent with the promotion and legitimation roles of the state. As already noted, an ideology comprises a core set of beliefs with associated principles to guide action. It provides a world-view, a framework for living, and it is assimilated by individuals as part of their socialization. In many ways it is comparable to the ideal-types that people use to organize their daily lives, in that it is a simplified picture of the world created to facilitate living in it. It differs from those ideal-types, however, in that it is concerned with abstractions rather than concrete instances, and also in the way it is moulded.

Ideologies are the ruling ideas of societies. They develop through social interaction in places, but their development and maintainance is closely controlled by the ruling forces in society. Thus all societies have an ideology which promotes the dominant class interests: under feudalism, for example, the divine right of kings was such an ideology in some places; under capitalism, it is the necessity of accumulation. Such ideologies are presented as representing everyone's interests, however, so that in Britain the necessity to accumulate, to have successful free market operations in which profits are made as a reward to enterprise and risk-taking, is advanced as in the interests of all. The major beneficiaries, of course, are those who reap most of the profits, so that the class goal is being presented as a general goal. To this extent, therefore, the ideology distorts reality, presenting a view which is not entirely consistent with the mechanisms that are being advanced. The goal of emancipation is to unmask that distortion, to reveal the ideology for what it is, a set of ideas constructed to promote a particular mode of production and its local interpretation. The ideology promotes a false consciousness by focusing attention on the empirical; emancipation removes that, by exposing the real.

The goal of emancipation is not simply promotion of theoretical awareness; it is to free people from the hegemony of ideology and to give them greater control over their lives through the revelation of how their societies really work. From this should come constructive social change, engineered by people fully aware of their situation. Emancipated individuals are those in full control of their own lives, not slaves to a structure, however tolerant that structure may be in terms of allowing choice and however successful it may be in providing material benefits. Thus the theoretical understanding that this form of applied geography provides is to be used not simply to deepen the awareness that could be gained through greater mutual understanding but also to provide a practical guide for the reconstruction of society.

Applied geography in this context is directed towards the general interest, not in the sense advocated in capitalist ideology but rather in a constructive critique of that ideology and all that it represents. It promotes the general interest by generating the conditions in which people themselves will decide what it is, not have a definition imposed on them. Human geographers' involvement requires creating what Harvey calls (1984, p. 7) a 'common frame of discourse', a means for understanding through which 'the myriad masks of false conflict be stripped away and the real structure of competing rights and claims be exposed'. There may appear to be no particular geographical role here, but

> There are many windows from which to view the same world, but scientific integrity demands that we faithfully record and analyze what we see from any one of them . . . The intellectual task of geography . . . is the construction of a common language, of common frames of reference and theoretical understandings (p. 7),

and thereby to show how space and place are integral parts of the continued restructuring of the world-economy. Indeed, space and place are used to promote capitalist ideology, to assist in the creation of false consciousness by implying that the world is divided into many independent parts. The empirical validity of that view must be challenged, not just through empirical analyses that show the many inter-connections between all parts of the global world-economy but also, and crucially, by theoretical analyses that reveal the oneness of the underlying mechanisms. Space and place do matter, but as resources to be manipulated, not as independent variables.

QUO VADIS?

Applied geography is almost invariably associated with technical control: the applied geographer is a technician, not a provocateur let alone a catalyst for radical change. This definition, as illustrated here, depends on a particular view of the interests that applied work is to serve. Technical and social engineering are promoted as being in the general interest but, as shown in this chapter, the general interest is ideologically equated with class interest and, as shown in chapter 2, what is in the general interest in one place may well be against the general interest in another. Mutual understanding is presented as in the general interest because it creates conditions in which tensions will be relaxed; but it neither tackles the sources of those tensions nor seeks to remove them. Emancipation is presented as in the general interest because it promotes real understanding by all, the unmasking of ideology. It tackles the sources of the tensions by demonstrating their origins and creating the conditions in which people will use their knowledge to remove those origins and create a new, non-exploitative, non-alienating mode of production in its place.

Any approach to applied geography cannot be neutral, therefore. It involves a choice and that is a political choice regarding how knowledge is to be produced and used. Further, it is not a choice that is unconstrained. In some countries, notably those of eastern Europe (see the relevant chapters in Johnston and Claval, 1984), the practice of geography is determined by the state's identification of national goals. Elsewhere, in the liberal democracies the main centres of geographical work are the universities, polytechnics and colleges that adhere to the concept of academic freedom, which implies that geographers themselves choose how they will practise their discipline. Nevertheless, the geographers are not self-employed; they must act within a constrained environment largely structured for them by the state, which is also the architect of national ideology. Direct interference is rare (though recall that in the late 1970s the government of Chile dismissed all non-monetarist economists from its universities) but indirect pressure is sometimes great, especially in periods of recession and restraints in public spending. As Peter Taylor (1985b) has demonstrated, geography only flourishes through patronage – either state or private – and that patronage has to be 'earned' by winning favour with the patrons. The need to demonstrate an applied expertise is thus especially important in periods of recession. In a booming economy, accountability will be less, and 'pure geography' is better able to flourish.

The present recession is clearly a period for an ascendancy of applied geography, serving the interests of the state and those which it supports. Of the three types of applied geography outlined here, technical control is by far the most acceptable, since it advances the promotion of accumulation. Mutual awareness is both acceptable and needed, since through communication it promotes appreciation of how societies work. It is part of the legitimation role of the state (especially with regard to matters within its own borders), and the promotion of its ideology, as Sir Keith Joseph (1985) clearly recognized when addressing British geographers about the relevance and value of their subject:

> Pupils need . . . [an] understanding [of] what is significant about a particular spot on a map or globe and why, as well as knowing where it is. It means understanding why particular changes are taking place and what their effects are likely to be, as well as knowing that they are happening. (p. 3)

He went further, however, in saying why geography should occupy a place in the school curriculum:

> Pupils can be helped to consider the application of democratic values and the rule of law and the consequences of their absence. (p. 6)

The awareness to be promoted is ideologically laden. The emancipation that would be advanced through the third type of applied geography will not be encouraged, however. It is unlikely to be suppressed directly, but the lack of encouragement, and the denigration of radical ideology and goals, would be sufficient to ensure that it did not flower. (And it might be studiously ignored, as is the case with 'radical' geography in South Africa.) This, after all, has been the fate of sociology in recent years. As Short (1984) overstated it: 'criticism of sociology is *de rigueur* for Conservatives who want to get on in public life' (p. 103). Today, 'pure sociology' is in decline, but 'applied sociology as technical control' (i.e. social work, social policy, and social administration) is acceptable.

The choice set for geographers is a hierarchical one. The initial choice is between either politically and ideologically acceptable or unacceptable applied geography, between technical control and/or mutual understanding on the one hand and emancipation on the other. The former is clearly the popular choice, and for many reasons – not the least of which is self-interest. Within it there is the sub-choice of technical control or mutual understanding. Again, the former is both the more popular and the more likely to win patronage, notably in terms of research support. It has many implications for the practice of geography, not least its emphasis on technical training rather than liberal/general education.

To choose applied geography as emancipation is to choose the geographical wilderness, it seems. Such activity, with its potential threat to the social consensus, is unlikely to succeed because it must counter the strong, well-supported ideology of the other two applied geographies. It may succeed in emancipating a few geographers, encouraging them to realize that to act as either technician or provocateur is to underpin that ideology and the mechanisms that it supports, a task that they may decide is unacceptable: if we believe that under capitalism (Johnston, 1986a),

> a problem solved somewhere is a problem created elsewhere, that all winners are compensated by losers, then to retreat to image processing, factor analysing and phenomenological contemplation is to be irresponsible. We need to process images, to analyse factors, and to contemplate – but to a purpose. Without that purpose, we are part of the problem, not the solution.

In the final analysis, the choice must be a personal one, and it poses a moral dilemma. An emancipated geographer is one convinced that radical social change is needed, and is tied to a manifesto aimed at winning the hearts and minds of people by emancipating them, in turn, from the negative ideology of technical control and preventing them from falling into the trap of believing that individuals can create their own worlds; only collective action can succeed, through a long-term programme of political education.

But what happens while this programme is being undertaken? Presumably the inequalities, which are the empirical outcomes of the mechanisms the applied geographers want to reveal, will continue to be reproduced, with many people suffering acute deprivation and premature death as a consequence. (Premature death can be expressed in the 'neutral' language of capitalist ideology as 'low life expectancy'. In realist language it could be expressed as the outcome of 'structural violence', the exploitation of some to advance the life expectancy of others.) Presumably the negative stereotypes, the nationalistic clashes and the military posturing will continue, bringing the threat of nuclear annihilation ever closer. Is it morally acceptable to allow these to continue, in the hope that eventually an alternative will be forged? Can any individual be sacrificed today for tomorrow's generations? (see Silk, 1982. An analogy of Wallerstein's, 1979, to illustrate this dilemma is discussed in Johnston, 1981; see also Johnston, 1986d.)

7

Geographical Fixations

Our broad training in both physical and cultural systems and our appreciation of landscape change in the natural and human sciences give us perspectives and insight that are rarely found in other disciplines.

W.L. Graf et al., 1980, p. 281

Reduce all social relations to relations of space and it would be possible to apply to human relations the fundamental logic of the physical sciences.

R.E. Park, 1926, p. 13

Society has allocated responsibility for the study of areas to geography; this responsibility is the justification for our existence as a scholarly discipline.

J.F. Hart, Presidential Address to the Association of American Geographers, 1985

[Regional and historical geographers] are particularly qualified to undertake the field observation and field study of problems recognised in a more systematic way and to conduct field tests of generalizations arrived at through systematic study.

NAS/NRC, 1965, p. 61

Geographers appear to be perennial debaters over the definition of their discipline, its internal structure, and its inter-relationships with other disciplines. In particular, attention focuses on the inter-relations between human and physical geography. Are they separate disciplines, so that the author of *The Geographer at Work* (Gould, 1985) is justified both in ignoring physical geography totally and in equating human geography with geography? Should they be integrated? Is synthesis the proper role of geography within the division of labour? These questions are tackled here, within a wider discussion of the content and role of human geography.

From the outset, it is necessary to stress my view, developed further in the next chapter, that there is no 'natural' discipline of geography, and certainly no 'natural' discipline of human geography – nor of any other social science. Human geography was not discovered; it was created – in particular places, at particular times, strongly influenced by one or a few leading individuals, to perform tasks acceptable to (or at least not unacceptable to) those with power in the relevant society, as Peter Taylor (1985b) has made clear. Once created, it – through its powerful leaders – developed its own *raison d'être*. It was not free of constraints, of course, either of patronage or of state sponsorship, but it was able to develop relatively independently within society. This involved individuals and institutions (such as the Royal Geographical Society and the Geographical Association in the UK) promoting the discipline, and fashioning a niche for it in the educational structure; implicitly, if not explicitly, their promotion of geography involved campaigning against the interests of other disciplines, a tactic described by Goodson (1981, 1985). These campaigns were set in particular contexts, or 'contingently related conditions' to return to the terminology of figure 4.3, and so were responses to perceived opportunities and constraints.

It is this contextual approach to understanding a discipline (well described in the chapters by Berdoulay, Capel and Grano, in Stoddart, 1981) that justifies what many see as the trite definition that 'geography is what geographers do' (though not in conditions that are necessarily of their own choosing). There is no *need* for a discipline of geography, though there is certainly a need for at least some of the work currently done by geographers. (It all depends on your definition of need, of course; for a narrow-minded definition, see Clark, 1985.) But there *is* a discipline called geography, historically constituted, embedded within academic, institutional and educational structures, and continually being reconstituted. It exists, and will continue to; what, then, is its rationale?

The existence of geography as an academic discipline in many places should not bolster the belief in the 'naturalness' of geography, for definitions and practice vary substantially from place to place (as demonstrated in Johnston and Claval, 1984, and in review articles in the journal *Progress in Human Geography*). Here attention is focused almost entirely on the British context: this, of course, is closely linked to the North American, but there are important differences (not least in the role of geography in school curricula) and we must not forget those between the situations in the United States and in Canada. Discussion of an individual discipline can take place only within a framework provided by the constellation of other disciplines. All of these are human creations too (though some may claim that a few – for example, mathematics – are 'natural') and so the nature of geography in any place has developed out of conflict (however polite, subdued, implicit) with other such creations.

Some argue against any academic division of labour, certainly any division within the social sciences. Proposals have been advanced for unidisciplinary (i.e. the dominance of one social science, usually economics), multidisciplinary, interdisciplinary and transdisciplinary reconstructions of social science, but Eliot Hurst (1980, 1985) has argued against them all, and propounded a new structure based on historical materialism and thereby involving a de-definition of geography (and sociology, political science etc.; Harvey's 'people's geography' goes with them: Eliot Hurst, 1985). As he puts it, we need a position:

> that *de*-defines the disciplines as we know them . . . it is not a matter of rearranging the fragmentation of epistemological space, but of transcending all such boundaries as we know them to emphasize the unity of scientific practice. Historical materialism as the science of human society ranks alongside mathematical and natural sciences as the only tenable scientific practices. (1980, p. 13)

Alongside this idealistic position is the more pragmatic one, which defends the division of labour, not in any absolute sense but, for geography, with regard to two features.

1 The division provides a necessary subdivision of knowledge, without which progress is very difficult in the advancement of understanding. This immediately raises the problem of the chaotic conception (see p. 63; Taylor, 1986a), and also that of reductionism (p. 91). Clearly, the first problem is a crucial one, for fragmentation into unrealistic parts provides for counter-productive scholarship. The second is only crucial if a given academic division of labour is ossified. The recent history of science, at least, shows this not to have been the case, with new disciplines being invented to follow promising research lines and to promote certain interests. (This is an interesting feature in the light of Pickles's, 1985, apparent claim that each science is linked to a particular, phenomenologically revealed, essence.)

2 The division provides niches for scholars and researchers to work in. Each discipline is internally structured, and individual careers within it are largely governed by discipline-defined peer review (see Johnston, 1983a, chapter 1); it is a society that practitioners join and which, through its (written and unwritten) rules, provides them with a comfortable environment. It can, and sometimes does, encourage

self-satisfied isolation, untrammelled by external contact and evaluation, and the role models that it provides may be stultifying. (As the recent history of human geography shows, however, external contact is not negligible and the discipline has been redirected by people dissatisfied with its current practice – though not always without difficulty: Johnston, 1983a.) But again, this is a question of practice rather than principle.

It is the pragmatic position that is adopted here. Human geography exists, so how should it be practised? That question is explored by looking at four geographical fixations of recent decades.

THE ENVIRONMENTAL FIX

As already noted, the inter-relationships of human and physical geography have been the source of considerable debate over a very substantial period. The grounds of that debate have shifted very substantially, however, as the following brief categorization suggests.

1 *The quasi-deterministic approach*, including much traditional regional geography, presented the discipline as very much the study of environmental stimulus and human response. Its extreme forms were some variants of both environmental determinism and possibilism, but most occupied a 'middle-of-the-road' position. All were holistic, because they stressed the interactions between peoples and environments (frequently termed the 'man-land' approach to geography).

2 *The holistic/synthetic approach*, also linked to some forms of regional geography and based on the case – put, for example, by Wooldridge and East (1958) and widely accepted among British geographers at the time – that a discipline (geography) was needed which took the findings of separate systematic studies (inside and beyond the discipline) and put them together to create a description of the whole phenomenon complex. This, it was argued, needed both human and physical geography because without either the whole could not be appreciated. Hence, it was contended, geography straddles the arts and the sciences (the social sciences were rarely considered) and was the bridging discipline between the two.

3 *The technical approach*, linked to spatial analysis and certain forms of applied geography. In this, the main focus was on methods. The so-called quantitative revolution of the 1950s/1960s led to consider-

able, valuable, methodological cross-fertilization between natural and social scientists: in Britain, because of the presence of both physical and human geography, the links were within geography, as in the works of Haggett and Chorley (1969); in the United States, the almost complete absence of physical geography led to extra-disciplinary links with sympathetic natural scientists, such as the geologist W. C. Krumbein. The links were almost all forged through common interests in technical issues and — apart from a few odd attempts at developing homologies (e.g. Woldenberg and Berry, 1967), many in the context of general systems theory — there was very little substantive linking of the two fields.

4 *The environmentalism approach*, which stressed the interactions among people, societies and the physical environment. On the one hand, the physical environment is identified as a resource to which people respond and on which they act, both aesthetically and economically. On the other hand, society is one of the agents controlling the rates of environmental processes. Thus, it is argued, physical and human geography need each other: without the former, the latter is ignoring the fundamental resource base for a materialist society; without the latter, the former is ignoring the forces that accelerate, retard, and catalyse environmental processes.

Of these four, the last has the strongest rationale. The determinism underpinning the first has been discarded; the arrogance of the second has been repudiated; and the weak intellectual base of the third has led to exposure of the poverty of links founded in shared techniques (what price a University department of chi-square analysis?) but epistemological incompatibilities (Johnston, 1986b). It is therefore, with the environmentalism approach that the rest of this section is concerned.

Environmentalism: Integrating Physical and Human Geography?

The need for physical geographers to be aware of human use of the earth's surface and for human geographers to appreciate the nature of societies' physical resource bases have led some authors to claim the necessity of a discipline of geography which not only brings the two together, but integrates them. Thus, for example,

focus on the interaction between physical event and human response ... does provide geography, physical integrated with human. (Gregory and Williams, 1981, p. 52) and

> There has been too much fractionation in recent years . . . a sad state
> of affairs for a discipline the principle strength of which lies, in my
> view, in an integrative approach . . . physical geographers, biogeog-
> raphers, and human geographers must pull together in the common
> cause. (Orme, 1980, p. 146)

In particular, several authors have argued that this integrative activity
should focus on the topics of resource analysis and management. The
rationale for this focus was clearly put by Goudie (1982), when invited to
answer the question 'Should Human and Physical Geography go their
separate ways?'

> . . . not only are the pressures posed on the environment by man
> becoming greater rather than less as time goes on, but the pressures
> posed for man by the environment are also not necessarily diminishing
> . . . who is the more dependent on his environment: the Kalahari
> bushman who can find food and water in an hostile environment, or
> the city dweller who could only last for a few days if his supplies of
> food and water were to be disrupted? The world *is* still a hazardous
> place, man *is* affected by his climatic environment . . ., human health *is*
> related to environmental conditions, climatic change *is* a threat to
> human welfare, environmental influences *cannot* be denied. (pp. 25–6)

None of this will be gainsaid here. Nor is an argument raised against
Balchin's (1981) claim – except the sexism – that:

> Man interferes with the natural ecosystems at his peril and, until he
> learns to harmonize his activities with those of Nature, he can expect
> only a problem-ridden existence. (p. 265)

All that is debated here is whether this requires the integration of human
and physical geography.

As currently constituted, human and physical geography involve two
separate groups of scholars, each seeking to understand their subject matter.
They share a common history of being wedded to evolutionary forms of
thought, and a recent history of associations with empiricism and
positivism. Among human geographers, positivism has been relegated and
replaced by various other approaches, such as the realist advanced here.
Among physical geographers, positivism still has a strong hold, despite the
development elsewhere of a realist physical science. Nevertheless, even if

both were to adopt a realist position, the two would remain very different research pursuits, for reasons set out above (see also Johnston, 1986b). The physical, chemical and biological processes that are the foundations for the production of landforms, weather and vegetation complexes are givens; the structures of society are human creations, forever changing.

What, then, of the resource analysis and management research which supposedly bridges this divide between human and physical geography? An examination of the literature (reported more fully in Johnston, 1983c) has identified four types of work:

1   *Resource appreciation* covering (a) work by physical geographers presenting their subject matter (e.g. climates, water) as resources which societies exploit; (b) work by human geographers on that exploitation (e.g. the use of water resources); (c) work by human geographers on 'popular' conceptions of nature and how the physical environment works; and (d) work by physical geographers promoting the understanding of environmental processes before they are altered through, for example, land use changes.
2   *People in the physical environment*, covering the human impact on physical ecosystems and processes, and application of the understanding gained through environmental impact studies.
3   *Conflict over the environment*, focusing on conflicts within society over the conservation, preservation and use of environmental resources.
4   *Human demands on the environment*, with particular reference to possible environmental constraints on social trends (both long-term, as in 'The limits to growth' debate, and short-term as with the Sahelian drought and famine).

In virtually none of this work is there any integration of the separate research by human and physical geographers into the processes involved in either the creation of demands on the environment or the processes that create the environmental resource. Work in resource management is conducted either from the environmental process perspective of the physical geographer (as discussed in K. J. Gregory, 1985, chapters 6 and 9) or from the social demand perspective of the human geographer (whether positivist and based on neo-classical economics or realist and based on class conflict). The two groups meet in common subject matter, but do not combine their research skills.

Nor do they need to. As Andrew Sayer (1983) expressed it:

in human geography . . . we may be interested in the causes of flooding and this will often include 'social' as well as 'natural' events. But although floods may be the effects of social actions this does not make floods social . . . Understanding the social character of the actions which caused the floods . . . would not be essential for understanding the latter. (pp. 55–6)

As pointed out earlier, everyday life is acted out in terms of stereotypes of individuals and societies, models which structure our thinking. We have similar stereotypes of the physical environment, conceptions of nature (Sayer, 1979) which direct our interpretations of the physical environment. In terms of understanding how we live, it is necessary to understand those conceptions, as the work of Gilbert White and others on various natural hazards made very clear (Burton, Kates and White, 1978). In order to improve the quality of life, it is desirable to improve those conceptions, with regard to environmental outcomes; resource managers need valid models of the resources they are managing. But there is no need to understand the actual way in which floods are produced, any more than the practising demographer needs to know how the birth control pill works. It is the probability of the outcome that matters and no more – which is why you find no physical geography in White's work.

Ron Cooke (1985) has recently disagreed with this position, arguing against the exemplar contention:

that physical geographers interested in the impact of farming on soil erosion do not need to understand why it is that farmers are intensifying their demands on the land. (p. 43)

But his example of slope failure in Los Angeles does not support his case. He shows that the greatest volume of failure occurred away from the areas of highest potential, and matched the distribution of 'newer affluent communities, often of young, articulate professionals, who can effectively command local influence' (p. 46). But this tells us nothing of the reasons why such communities exist, and although knowledge of their attitudes to public help and the availability of insurance allows us to understand their activities Cooke's work in no way brings together the separate modes of analysis of contemporary physical and human geography. He is right in arguing that:

the natural processes that fashion slopes impose themselves intimately on the patterns of human behaviour, and the complex network of action and reaction demands that they, too, are understood. (p. 46)

This cannot be gainsaid, for appreciating the links between society and environment is a crucial task that geographers undertake. Cooke places such appreciation at the centre of geography, which he presents as a web of relationships between human activity and environmental contexts. He then mildly castigates:

> some extremists on the fringes of the web [who] will probably see little of interest in the centre of it. In writing the post-war history of geography, they will be tempted to ignore or underplay, for example, those geographers who have contributed most influentially to studies of the relations between communities, cultures and the physical environment. The names of Carl Sauer and Gilbert White, and their numerous students, will not figure prominently in their reviews. (p. 46)

But, in my reading of their work, I identified no physical geography in the writings of Carl Sauer and Gilbert White. They were interested in society's reactions to environmental processes that they took as given, and in how those reactions were reflected in the landscape. Cooke argues that:

> we should perhaps be seeking to understand, say, *either* how water moves in soil *or* what motivates people to exploit hills. We can, of course, choose either. But the problems in between remain, and I believe them to be fundamental, fascinating, and central to geography. (p. 46)

I agree. But, as Gilbert White's work shows, you do not need to understand how rainstorms and floods are produced to study how people react to the flood hazard.

What, then, is all the fuss about? In part, it is a problem of scientific politics, for people perceive that a divorce of physical and human geography would be deleterious to both. (Not all agree: see Worsley, 1985.) I see no threat, however, as laid out below and in the next chapter. Secondly, a lot of the fuss is nothing more than semantics. The term geography has two major usages. The first – the vernacular – refers to the features of places: physical geography is about landscapes; human geography is about what is done where. The second – the academic – refers to the content of the discipline: physical geography is about understanding environmental processes; human geography is about understanding why things are done by people where they are done. Unfortunately, few physical geographers know much about academic human geography, and vice versa. Thus when physical geographers are arguing for their integration with human geography they are using the latter in the vernacular sense. *As so often happens, our meanings are distorted by the words we use, both by us, and more importantly, by those who read them and interpret them according to their dictionaries, not ours.*

Divorce, Separation or A Compatible House-sharing?

So, am I advocating the separation of physical and human geography? Is my statement (Johnston, 1983a, p. 6) that: 'there is no apparent need . . . for an integration of the two fields as they are currently practised' the first move towards the divorce courts? No. There is much to be said for maintaining, indeed enhancing, the sort of contact currently practised, without any necessity for grandiose, unnecessary, integrationist claims.

Much of the basis for my position rests in the two definitions of applied geography, other than technical control, advanced in chapter 6. As an educational discipline, geography does much to promote awareness – of the nature of the physical environment among human geographers; of the nature of society among physical geographers; and of the various conceptions of nature among all. (It could well be argued that in universities the student's awareness is much greater than that of the specialized staff, for the physical:human divide is a major one with regard, for example, to seminar attendance.) Dave Pepper's (1983) discussion of the problems of stimulating such awareness provides valuable insights to this, and his book on *The Roots of Modern Environmentalism* (1984) develops the sorts of theme that are valid for both physical and human geographers. We need to be aware of the social construction of environmental images, and the sorts of behaviour that this results in (O'Riordan, 1977), and we need to be aware of the structural bases for our attitudes towards the environment (just as historian Keith Thomas, 1984, has made us aware of our attitudes to flora and fauna). As Sayer (1979) reminds us, nature is something appropriated for human use under capitalism; it is the focus of human labour, so that:

> Human labour, as distinct from animal labour, requires understanding of Nature's mechanisms, and this knowledge is not innate but socially acquired. Knowledge of Nature is never produced through an unmediated reflection of the subject upon Nature but always uses 'means of production' in the form of the existing social knowledge. (p. 29)

Part of the emancipatory task of geography, therefore, is to unmask the ideological presentations of nature as something that we control totally (recall the debates about nuclear power). We need to demonstrate, with Lukács (1923), that:

> Nature is a societal category. That is to say, whatever is held to be natural at any given stage of social development, however this nature is related to man and whatever form his involvement with it takes, i.e. nature's form, its content, its range and its objectivity are all socially conditioned. (p. 234)

An academic discipline which does that is, to me, a viable one, even if it does it in two separate parts.

THE SPATIAL FIX

From the mid-1950s on, a movement promoting geography (basically human geography) as spatial science – alternative terms were locational analysis and spatial analysis – spread outwards from its major foci at four centres in the USA (Johnston, 1983a). It drew on a variety of roots, some in geography (including work by Scandinavian and German pioneers) and some in other social sciences. As has been pointed out, there has always been a geometrical tradition within geography. But this movement – rapidly dubbed the quantitative and theoretical revolution – took a new approach, and was hardly a continuation of that tradition, which is now mainly in surveying and mapmaking. It was presented either as a replacement for, or as a logical development of (by introducing precision and rigour), the then dominant regional approach, which Peter Gould (1979) described as:

> bumbling amateurism and antiquarianism that had spent nearly half a century of opportunity in the university piling up a tipheap of unstructured factual accounts . . . it was practically impossible to find a book in the field that one could put in the hands of a scholar in another discipline without feeling ashamed. (pp. 140–1)

Alongside the criticisms of current approaches to human geography, spatial science was rationalized (retrospectively) as a necessary move given shifts in the nature of society. As Kevin Cox (1976) expressed it:

> In a world of spatial interdependence . . . locally experienced environmental dependencies lose their rationale: men relate less and less to the land on which they stand and more and more to

socially-created geographical patterns over a much wider area. (p. 193)

In a locally oriented society, environment is crucial: in a globally organized society, spatial relations are crucial, and the 'new geography' thus had to stress the latter. It comprised two elements. The first was *the focus on spatial distributions* – initially via deductive models but later, as their axioms were found wanting, through more inductive searches – with the aim of identifying what Schaefer (1953) termed morphological laws and laws of coincidence, generalizations about the spatial arrangements of individual phenomena and their correlations with those of other phenomena; human geography was to become a science, within the (implicitly positivist) general conception of that term. The second was *the use of mathematical languages for modelling and statistical procedures for hypothesis-testing*. Put the two together, and you get the orientation to human geography defined by Berry and Marble (1968) as:

> building generalizations with predictive power by precise quantitative description of spatial distributions, spatial structure and organization and spatial relationships. (p. 6)

Various approaches to the structuring of human geography as spatial science were adopted. Dick Morrill (1970), for example, focused on the location of all activity as a reaction to the frictions of distance: location decisions were taken so as to minimize the consequent transport costs, as were movement decisions (e.g. journeys to shop). Peter Haggett (1965; Haggett, Cliff and Frey, 1977) focused on the outcomes, identifying five (later six) main components of a spatial system, each subject to the operation of basic spatial laws. Others sought to specify the fundamentals of spatial science more precisely. In a pioneering, but largely disregarded, paper, Allen Philbrick (1957) identified what he termed the 'principles of area functional organization'; they hinged around the basic postulate that:

> Human occupance is focal in character (p. 303),

with the patterns of nodal organization which reflected this displaying hierarchical spatial structures. John Nystuen (1968) followed this with a search for the concepts necessary and sufficient to a 'geographical point of view'. He identified three: direction (or orientation), distance, and connection (or relative position). And Gerard Rushton (1969) used inductive procedures to identify patterns of 'spatial behaviour', the general rules that:

exist independently of the environment where the decision is made
(p. 393),

and can be derived from studying particular patterns of 'behaviour in
space'.

What these approaches shared was a desire to show that human
geography is as rigorous as the other social sciences, most of which were
then in a purple patch of quantification, and also that it had a particular
perspective to contribute to the social sciences. That perspective was the
focus on space – on the spatial variable. It was very largely a restricted view
of space, however, for it focused almost entirely on relative (and relational)
space, on distance as the determinant variable in location and movement
decisions; what happened in a place depended on its relative location. In
terms of the Haggett system, the earth was a surface. Absolute spaces were
identified – through the equation of regionalization with classification – but
only as simplifications of that surface for didactic and analytical purposes.
(According to Haggett, like the NAS/NRC committee (1965) quoted above,
regions were the 'test beds' for systematic ideas.)

This approach has been interpreted since as based on a spatial fetishism.
Space and/or distance were reified; inanimate things were given an existence
of their own and a power to influence, if not determine, people's actions.
Distance is, of course, an important influence – as is made very apparent in
Geoffrey Blainey's (1966) portrayal of Australia's development as a
consequence of *The Tyranny of Distance*. But it is really only a constraint to
the extent that we let it be, because we have built societies that use distance
as a barrier, as a form of cocoon. The tyranny of distance has been
incorporated into society, and used by it, so that relative space is really what
we make it and has no independent existence (as Bob Sack argued
powerfully in a series of articles in the early 1970s; see Sack, 1974, and
Sayer, 1985); and of course, as illustrated in chapter 3, societies have also
been concerned to annihilate space by time, to get rid of the tyranny and so
foster the process of capital accumulation.

The argument against the fetishism involved in spatial science, because of
its focus on relative distance and its naive belief that a separate social science
can be built on that reification of space, is important; even more important
is the relative absence of consideration by spatial scientists of absolute
space. This is surprising because our daily lives are lived out in absolute
spaces, and our societies are organized similarly. We live in nested, partially
overlapping containers, some of which have more porous walls than others.
For most individuals, there are four such containers – home, workplace,

local government and state – each of which has clearly defined boundaries, most of which are in some way 'policed'. (The boundaries of local governments are the general exceptions to this; though see Johnston, 1984d.) None of these is 'natural'; each is created, defined by us, in our local context. (Places matter in the interpretation of places.)

We live out our lives, in large part, in a system of absolute spaces. These are created, at least in part, to manipulate relative space and the frictions of distance. We have separate homes, with walls, to provide us with privacy and separate us from contact with others. Some of us have gardens, too, to provide a cocoon and a further layer of insulation, especially effective if they are surrounded by hedges/fences (and with assumed unfortunate consequences when they are absent: Coleman, 1985). We have states to protect us from others. Many of our institutions are organized in terms of absolute spaces as well, as David Harvey (1974b) illustrated for the Baltimore housing market. We have ghettos, into which we confine our 'undesirables', and we have asylums and prisons to contain those we perceive as threats; nowhere is this clearer than in South Africa, where the apartheid system involves the careful manipulation and policing of absolute spaces to achieve control by a racial minority.

Many of these absolute spaces are legally defined and have demarcated boundaries. They are part, as Bob Sack (1980, 1983, 1986) has told us, of a territoriality that is basic to the organization of societies. To some, this territoriality is innate – one of the essences that phenomenologists seek (p. 55) – to humans, as to other animal species. Sack does not take this view, but rather just notes that:

> social organisations are often territorial ... the assertion by an organisation, or an individual in the name of an organisation, that an area of geographic space is under its influence or control. Whereas all members of social organisations occupy space, not all social organisations make such territorial assertions. The social enforcement (and institutionalisation) of such assertions ... provide the context necessary for social facts to exhibit the first type of spatial properties. The forms such territorial structures take and the functions they provide depend on the nature of particular political economies. (1980, pp. 167–8)

Thus absolute spaces are not natural; they are human creations, produced in a particular context, to serve a particular purpose, and once produced are an element of the milieux within which socialization occurs.

The systems of territorial containers – both *de jure* and *de facto* (e.g. the 'turfs' of neighbourhood gangs) – are in no way fixed; they are continually changing. The state system is in constant flux, for example, as those in power seek to extend the territory over which they have some power: to advance accumulation (colonialism, for example, and the creation of the EEC); to promote legitimation (the Argentinian invasion of the Falklands/ Malvinas?); and to create social cohesion (Israeli expansion?). Others – both other states and would-be states (e.g. nationalist movements) – compete with them, occasionally producing armed conflict over the possession of certain areas. Some of the contests are culturally-based, some have very vague roots in the visions of grandeur of certain individuals; but many have economic roots (colonization is the basis of Ulster's problem, for example) or goals (Scotland's oil).

Relative space, or distance, is a hindrance and has to be overcome; the process of annihilating space by time involves restructuring the pattern of relative locations in order to promote individual and group gains. In the development of telecommunications and the delivery of weapons, space has now been virtually annihilated; messages can be relayed almost instantaneously to any major settlement, and nuclear weapons can be delivered anywhere in a matter of minutes. These twin 'developments' have been crucial to the extension of capitalist hegemony. Power has been sustained by transforming the map of the world. It has also been sustained by the creation of absolute spaces, whose maintenance and restructuring is similarly crucial to the exercise of economic and political power.

A Valid Point of View?

Space is central to the organization of societies, but there is much more to its influence than the friction of distance that has been the focus of spatial scientists. By concentrating on relative space as a fixed variable they have both (a) ignored its creation and modification as a resource by societies and individuals and (b) paid insufficient attention to its manipulation into absolute spaces.

Given the broader focus implied by the previous paragraph, is the spatial fix a viable focus for a separate social science of human geography? As already stressed, Sack has provided a series of authoritative statements arguing against a separate discipline that feeds in the spatial variable as an autonomous influence. Geometry cannot provide an explanation, only a description; to explain you must appreciate the context of that description, for you cannot separate space from substance, any more than an historian can separate time from substance and create a discipline with time as the independent variable. (Note that some disagree. Coffey, 1981, for example, has argued that 'a non-relational concept of space is not only a useful one for conducting geographical inquiry, but also one which is in accord with the economy and generality which characterize science': p. 42.) Space has no independent existence outside its construction by society, any more than nature (or vernacular physical geography) has. Neither is irrelevant, and because of either one or both a human project may fail, with the failure leading to a rethinking of the nature of the constraint (assuming that it is not treated as an anomaly: p. 15).

Spatial components characterize most human projects. In many, they are so trivial as to be virtually irrelevant. In others they are crucial, and thus non-trivial. In the social sciences in general, therefore, it is necessary for proper attention to be paid to those non-trivial spatial components. To separate out the spatial – whether relative or absolute – and study it and it alone is undoubtedly to commit the sin of studying a chaotic conception. But to ensure that the spatial is neither overlooked nor torn out of context is to make a valid contribution. A human geography that makes such a contribution is viable, therefore, so long as its practitioners are providing a perspective from *within* the social sciences rather than *for* them. In this way the general understanding of society will be promoted, as the study of a whole not a set of disparate parts.

THE REGIONAL FIX

The regional fix preceded the spatial fix as the geographer's major fetishism. There were many interpretations, but fundamental to most was the belief that the earth's surface is divided into a mosaic of different regions each of which had a particular character as a consequence of the symbiosis between the physical environment and the society occupying it. Hence they were organic wholes, 'natural regions' which were not to be confused with the artificial regions imposed on the earth's surface by the drawing of political boundaries (Johnston, 1984a). As already noted (p. 125), this approach was imbued with a strong, if implicit, environmental determinism.

The term region was not discarded when the regional fix was replaced by the spatial. Instead it was modified, and used by some spatial scientists to legitimate their quantitative work; they could define regions 'objectively' and could model them as systems. This 'regional science' replaced humanistic appreciation by quantitative, pragmatic description; the study of places all but disappeared.

For a while – perhaps 25 years – regions were little more than heuristic devices for most human geographers. A few continued to argue for the understanding of place as an important geographical contribution, particularly places other than one's own. But the preaching was rarely satisfied in the practice, and regional geography went into abeyance. And then, in the late 1970s, people started to ask if they could have it back. Derek Gregory (1978) ended his exposé of the ideological bases of the many varieties of human geography by noting that:

> Ever since regional geography was declared to be dead – most fervently by those who had never been much good at it anyway – geographers, to their credit, have kept trying to revivify it in one form or another (p. 171),

and then introduced his own try with a call for knowledge of

> the constitution of *regional* social formations, of *regional* articulations and *regional* transformations . . . it's not difficult to point either to the warrant provided by geography's long-standing commitment to places and the people that live in them or to the regional structures which persist in contemporary space-economies.

It is probably fair to say that many readers were uncertain then how to interpret this; I doubt if they are now.

The case for the sort of regional geography that Gregory was calling for is integral to the arguments that penetrate the whole of this book. Quite simply – places matter. The capitalist world-economy operates a system of imperatives globally, but the implementation of those imperatives is undertaken, in large measure, by people socialized in particular milieux, each in a cultural region with its own language, values and characteristic attitudes. Like capitalism itself, those cultural regions are not fixed, for they are being constantly restructured. But they are interpretations of how to live, how to interpret nature, how to respond to the tyranny of distance. Thus the United States, Australia, Mexico, France, Israel and Singapore are all capitalist countries, but the empirical appearance of capitalism varies among them because of the ways people have created different modes of living within that mode of production.

In large part, this place-to-place variability (in may cases as important *within* as *between* countries) is a result of history, of the construction of a capitalist social formation on pre-capitalist foundations. The latter varied enormously, because of their relative isolation; unlike capitalist societies, pre-capitalist ones have been largely unsuccessful in annihilating space by time. Even minor differences between neighbours, perhaps one the offshoot of another, tended to magnify over time, producing a complicated mosaic of cultural variations – in dialect if not language, in food tastes, in land tenure and farming practices etc. – which capitalism has failed to erase, despite the great mobility of population and the introduction of homogenized national and international cultures (the so-called coca-colonization of the world: see Peet, 1982, 1986). Indeed, the development of industrial capitalism not only to some extent absorbed such cultural variations, it also created more, as in the attitudes to married women working displayed in different UK industrial regions (McDowell and Massey, 1984; see also Langton, 1984). Capitalism has its history, too, which is deeply embedded in the places created and recreated in response to its imperatives.

So-called 'traditional' regional geography lost favour because of the onslaught that it faced from the 'quantitative and theoretical revolution', an onslaught against which it was powerless because of the poverty of its empiricism and its continued links to an explicit environmental determinism. Some still adhere to it, but their attempts to promote a revival are based on weak rhetoric. Fraser Hart (1982), for example, argued that good regional geography contains 'evocative descriptions that facilitate an understanding and an appreciation of places, areas and regions'. (p. 2) The region to be described is 'a more or less homogeneous area that differs from other areas' (p. 9), and no standard definition can be offered:

> Regions are subjective devices, and they must be shaped to fit the hand of the individual user. There can be no standard definition of a region, and there are no universal rules for recognizing, delimiting, and describing regions . . . Understanding is more important than classification, and the core usually is more important than the fringes. (pp. 21–2)

Three key themes should recur in regional study: a sense of time, of continuous change; the relationship between scale and detail; and 'the importance of the physical environment, which provides the stuff that people have to work with, and sets the stage on which the human drama is played' (p. 24).

Although I can agree with the general manifesto, I cannot accept the empiricism with which Hart wants to put it into action. Nor am I sure, despite his urgings, that his sort of regional geography is what society wants of us. Certainly, we should be aiming 'to satisfy human curiosity about how much of what is where and why it's there, about the where and why of places and people, about the land and how people have used and abused it'

(p. 19). But to offer no more than superficial glimpses (however well illustrated by however many slides: de Souza, 1983) is to do little more than the Victorian pioneer explorers who returned to the great geographical societies with their lantern slides and tales of exotic cultures. Clearly we must *sensitize* people to the variability of places,but we must help them to appreciate the sources of that variability; to use Hart's own term 'our obligation to society' is to promote emancipation, not just awareness. Regions must be presented for what they are: human creations of the contingent conditions within which the imperatives of the mode of production are acted out and etched into the culture and landscape of places.

Places matter: regions matter. They are the milieux in which people learn how to survive. Our obligation is to understand them, to show that they are forever changing, and to indicate why. (See the recent collection of essays by Gregory and Urry, 1985.) We must do more than satisfy curiosity; we must build useful knowledge.

THE SYSTEMATIC FIX

The decline of regional geography in the 1950s and 1960s was associated with the ascendancy of systematic geography. For decades previously, geographers had argued that the two were symbiotic, with the systematic geographies(ers) providing the parts that were assembled into wholes by the regional geographies(ers). But in the 1950s and 1960s the assembly line was almost deserted; people became experts on the parts alone. Most common was the division of geography into topical specialisms. The major split was between physical and human. Within the latter, there was a two-tiered division. At the top level were economic, social, political and urban, and each had its subdivisions – industrial and agricultural within economic, for example; inter-urban and intra-urban etc. Most human geographers associated themselves with one of the major divisions, if not one of the subdivisions, and the discipline became both compartmentalized and outward-looking. The members of the various compartments found that they had little in common with most of their colleagues (certainly those with whom they worked on a day-to-day basis in a university department, except perhaps with regard to methodological, especially quantitative, issues) and much more with the members of other disciplines. (That external orientation rarely became symbiotic, and few economists, sociologists etc. found much stimulus in and use for the work of geographers.)

Not all human geographers entered one of these topically defined compartments. Some, especially historical geographers, retained a broader perspective. Others – the true spatial scientists – focused on spatial arrangements, irrespective of the substantive subject matter; Peter Haggett, for example, has focused much of his research on spatial spread and diffusion, using similar models and techniques for the analysis of both measles and unemployment. This was spatial systematics – the study of points, networks etc. – rather than topical systematics.

This fragmentation merely exacerbated the problem that the academic division of labour in the social sciences suffers from; the whole – the global world-economy – was taken apart so that the pieces could be investigated, but nobody put it together again. To study the individual social sciences introduces the possibility of focusing on chaotic conceptions; to fragment them even further changes the possibility into a high probability. Lip service to systems approaches suggested that a framework was available, and would be used to assemble the parts at the proper time, but there was little evidence of what the end-result might be.

The development of a realist approach countered this tendency with, for example, David Harvey's (1973, p. 15) statement that issues of distribution cannot be separated analytically from those of production: economic and social geography are integral parts of each other. Understanding how capitalism operates needs theory which incorporates social (the relations of production) as well as economic (the forces of production) factors, and which recognizes the central purpose of politics (the state) in binding them all together. Work on the geography of development and underdevelopment at one scale and of residential patterns in cities at another helped to make this point and to demonstrate the intellectual futility of much of the compartmentalization (as, for example, in the debate over whether there is anything that is specifically urban; see Dunleavy, 1981).

IN SUMMARY

The fixes that I have discussed in detail here can be summarized in four simple questions:

1 Should human and physical geography remain together as an academic discipline?
2 Is there a valid spatial scientific contribution that human geographers can make to social science?

3 Is there a necessity for regional geography?
4 Should human geography be practised through a series of topical systematic subdisciplines?

My answer to them all is yes, for I have no desire to suggest even the partial closure of any doors passage through which may aid understanding. But in each case my 'yes' is tempered by arguments against excesses, which unfortunately have been produced all too often in other people's answers and in reactions to them.

With regard to the human-physical link, I have argued that this is necessary to promote emancipation from the ideological portrayals of nature as well as awareness of the environmental constraints to social action. The bridge between the two must be strong. But there is no need for the promotion of an integrated geography, a discipline that synthesizes both natural and social science. Indeed, that is promotion of an impossibility, because of the epistemological differences between the two, which I see as irreconcilable.

Turning to spatial science, my 'yes' is tempered by recognition of the need to integrate the study of the spatial variable with the entire body of social science and the need for much greater attention to the creation of absolute spaces in the structuring and restructuring of societies. Within the social sciences, the value of a discipline that highlights the spatial elements is, to me, undeniable, but this is neither to claim exclusive rights to that perspective nor to believe that the social sciences might not evolve (in some places?) in ways that remove the need for an identifiable spatial point of view. Linked to this is my positive response to the question on regional geography. Again, it appears to me that the case for the focus on place (really an extension of the focus on absolute space) is undeniable. But it must be neither empiricist/positivist nor romantic/idealist; places must be studied for what they are – human creations of some form of order within which life can be lived and structured. Here, too, is a major contribution that geography, as currently constituted, can make to social science. Part of that contribution may relate to particular topics – to industry, offices, agriculture, or infant mortality perhaps. Concentration on such topics in research programmes will advance their understanding, but only if conducted in a holistic framework. Even if we avoid the problem of chaotic conceptions we get nowhere if our rational abstractions are not returned to the whole.

Geography has always been a broad discipline, and its breadth has led many of its practitioners to seek a particular focus for the discipline, that can be presented to others as its *raison d'être*; the need for such a *raison d'être* is always felt most keenly in a period of recession. But in my view all such searches are in vain, because they inevitably lead people either to promote a relatively narrow view of the discipline or to make such arrogant claims for breadth that the special pleading is rapidly rejected. Breadth is strength, because social science deals with a massive integrated subject matter. Specialism within that is necessary for research, but not blinkered specialism. Human geographers, as presently constituted, bring spatial (in the broadest sense of that word) and environmental perspectives to the social sciences. They must continue to do so: as social scientists who are geographers, not as geographers who are social scientists.

8

A Human Geographical Education

There is a stark choice to be made: pursuing the arms race, or moving to a more sustainable world political and economic order. Geographers must work to break the silences of the world, to counter collective mania with reality and hope.

Sir Shridath S. Ramphal addressing the GA/RGS/IBG, 1985

Human geography as an academic discipline has no direct link to a profession; very few people practise as geographers and there is no 'natural' outlet for the graduates of its university, college and polytechnic departments. For this reason, the discipline occupies a very uneasy position within educational structures: defining its role and maintaining/enhancing its status are issues of concern for all geographers, especially at times when the nature of education is under close scrutiny and the resources made available are limited. This chapter looks at some aspects of that role, as it relates to the approach to human geography outlined here; the focus is almost exclusively on the British situation.

Until about the mid-1950s, the academic position of geography was relatively secure in further and higher education. As the result of successful promotion and lobbying early in the century (notably by the Royal Geographical Society and the Geographical Association), geography was accepted as a major discipline within the British school system, and the development of the academic discipline in the universities very much reflected this. The main purpose of the small (in number and size) departments of geography was to prepare graduates for careers as teachers – though not to train them for that role, for the provision of the academic foundation was almost invariably separated (in time and place/personnel) from the vocational training element. The task of the teachers was then to impart a general geographical education to virtually all secondary school pupils, and to prepare a small number of them for entry to universities, from which they would in turn graduate as teachers. Thus geography was perceived very much as part of general education, promoting awareness of environments and peoples, at home and abroad, and the academic discipline was geared to that end. Because they were in universities, the academics were encouraged to do research; the amount done was relatively small.

From the 1950s on, this cosy, closed system was changed as the result of a number of interacting factors. The expansion of higher and further education meant more students and, since geography was a large, popular subject at schools, the demand to read for geography degrees grew. There were more academics as a consequence, employed in universities (and later polytechnics and other institutions) where a greater emphasis was being placed on research – and where personal advancement was closely linked to research records. At the same time, the pro-science ethos saw the growth of systematic studies, at the expense of regional geography (the staple of the education-oriented syllabuses), and the beginnings of the positivist revolution (Johnston, 1983a; Johnston and Gregory, 1984).

No longer were the academic discipline and the school subject tied together in mutual symbiosis, though a large number of graduates still entered teaching as a career. Geographers were finding other outlets for their knowledge, particularly in the fields of spatial planning (urban and regional, town and country etc.). The foundation for this rapidly expanding activity was survey of what was done where, especially what the land was used for, which fitted well with both the empiricist orientation of much academic geography at that time and its strong environmental underpinnings. The work of Dudley Stamp on land use and capability received wide official recognition, and it was through his efforts and those of a few others (Beaver, Willatts, Wooldridge, for example) that not only was geography accepted as a useful discipline but also geographers were able to fill senior positions in the planning hierarchies. The particular skills of local fieldwork (careful observation, recording, and presentation, accompanied by discursive analysis), developed by academic geographers to promote the educational role of their discipline, found other major outlets.

With the development of spatial science in human geography, the academic discipline enhanced the analytical contribution that it was able to make to planning activities – and no doubt the close links (professional and personal) between geographers and planners helped to ensure that those contributions were in demand. Human geography developed as the academic discipline whose applied outlet was in the field of spatial engineering. (Physical geography developed in the same way, though somewhat later, with respect to environmental engineering.) In this way, the research and teaching activities of academic geographers became increasingly divorced from the discipline as it was taught in the schools: there, the awareness goal of general education remained dominant. This produced tensions: what sort of geography at schools for what sort of geography at universities, polytechnics and colleges? Those tensions underline much of what is discussed in this chapter.

The link between academic geography and town planning remains, but is now much weaker, especially in terms of recruitment of geographers into planning. In part this was because of the professionalization of town planning itself: it now has its own degree courses (mostly postgraduate, with geography graduates major recruits) and professional institutions. More importantly, by the mid-1970s its period of booming growth was clearly over. For academic departments of geography, therefore, other outlets for graduates had to be found. If the discipline were unable to 'sell' its graduates on an increasingly tight labour market, then it would not attract students — a crucial issue since in large measure the number of academic jobs available is a function of student enrolments. Geographers had to ensure their own future, by preparing for that of their graduates.

The career-oriented outlook is seen as necessary in order to ensure the vitality of the academic discipline: human geography must be 'relevant' to the demands of the world. Academics needed to tailor their offerings accordingly. Concern about output was only part of the issue, however. There was also concern about input, for the contents of academic geography and school geography were becoming more and more apart: the former was promoting utility and positivism; much of the latter was still promoting awareness and regional empiricism. Should the two be brought together again? How? And which should yield?

Given the absence of a guaranteed market for their graduates, therefore, academic geographers (human and physical) are much concerned about the content and the image of their discipline: their careers depend on it. So, too, are school geographers, though for somewhat different reasons. Thus all geographers are engaged in political battles, promoting their discipline externally while debating its nature internally. The remainder of this chapter looks at those battles, from three perspectives.

HUMAN GEOGRAPHY IN HIGHER EDUCATION

The goals of a degree programme in geography are rarely stated very explicitly, and the same is certainly true of individual courses within those programmes. Institutions advertise their offerings in terms such as

> our courses offer opportunities for you to develop practical skills in field, laboratory, library, archive and computer settings.

> Study environmental and social problems in an informal atmosphere. A sound professional education giving direct entry to stimulating careers.

and

> a wide choice of Human and Physical options, plus subsidiary subjects, which develop skills of collecting, analysing and presenting information about Man and his environment in order to provide an education suitable for many careers.

> (Statements taken from advertisements in *The Geographical Magazine*, August 1985)

Virtually all potential degree-level students will have studied geography for the last five years of their school careers; it is assumed that they know what geography is and have defined their goals for a degree in geography, so all that is needed is to emphasize the nice place, choice, and career-orientation. University, polytechnic and college departments have adopted the consumer sovereignty model, and orient their offerings to what they believe the students *want*. No doubt in general the academics believe that it is what the students *need*, as well, but survival depends on filling places.

One can speculate unendingly about the reasons (a) for the acceptance of the consumer sovereignty model, and (b) for the lack of very explicit degree programmes. With regard to the latter, for example, it may well be that it is the only way to obtain consensus among a very disparate group of academics, with very different definitions of science, applied orientations etc. Educational research has shown, for example, that most university departments (in all disciplines) are far from homogeneous, and that their members' main academic contacts are within the 'invisible colleges' of like-minded researchers in other places. Their reference group of peers is not the people they work with daily (Knorr-Cetina, 1982; Trow, 1976), so it is perhaps not surprising that they lack a collective, departmental view. This is perhaps especially so in most departments of geography in which the main goal when making appointments is to broaden the range of subjects covered rather than to develop specialist research teams. Each department is thus a disparate body of individuals, working as individuals. They perceive the need to present a departmental image, but do this by stressing breadth and choice rather than cohesion, by promoting themselves in general rather than in specific terms. And yet in many the breadth is constrained very largely to a particular view of geography.

Just as many academic staff are not very precise at defining the goals for their degrees and courses, let alone spelling out the curricula to be followed (exceptions to this in Britain are those whose degree schemes are validated externally, notably by the Council for National Academic Awards), so students are not necessarily the rational buyers assumed by the consumer-sovereignty model. Many select geography as a degree subject because they liked it at school (often reflecting as much on teacher as on content) and because they performed relatively well at it. The choice of where to apply to is often based on very partial information, employing no clear criteria with which to evaluate the various offerings. And once in a degree school, the choice of optional subjects is often little better than haphazard. Geography – and perhaps human geography more then physical geography – is not necessarily a subject taken by drifters, but it is one which many select by default rather than by positive choice. Many are not very clear what they expect from a degree in geography – either before, during, or after.

Within most degree programmes, the implicit goals are those of the three types of applied geography outlined in chapter 6: technical control, mutual understanding and emancipation. Of the three, the first (implicitly) receives by far the majority of attention, reflecting not only the dominance of empiricism/positivism among human geographers at present but also the ideology of the dominant forces within society and the general demands for education-cum-training that is relevant, both to the solution of current societal problems and to the preparation of graduates for careers. Given the absence of a major career market – geography is not a vocation – then students must be given the basis for competing in a wide range of job markets. Human geographers are clearly not alone in this, for careers advisors report that nearly half of all employers seek graduates without specification of the degree subject. They want educated people, and will provide the on-the-job training themselves; but in that context what sort of education should a degree in geography provide?

In promoting their degree courses human geographers emphasize that within the context of disciplinary interests they provide a range of general, transferable skills. Indeed, it is usually the skills element of the programme that are compulsory. Thus all undergraduates are practised in the application of literary skills, in the written and spoken presentation of materials. This focuses very substantially – in some places almost entirely, and obsessionally so – on the writing of essays: most degree results are based not just on the ability to write a structured piece of prose, but to do it in about 40 minutes, without any recourse to reference material, responding to an unseen question. The ability to react almost instantaneously to a literary stimulus is, to many, the *sine qua non* of an educated person. And that stimulus may be obscure, ambiguous, contorted: it is all part of the test. But the testing of that skill in other contexts, let alone the testing of other skills, is resisted by many: instant recall, and the ability to structure relevant material in a very stressful situation, is still the prime test of an undergraduate's ability.

Alongside this, human geographers increasingly promote numeracy as a necessary skill for their undergraduates – though ability in it is much less comprehensively tested. Virtually every first-year undergraduate course now contains a compulsory element on quantitative methods, and most second-year courses do too: in addition, many provide instruction in computer use, if not programming. This is justified, validly, on the grounds that: quantitative analysis is central to so much contemporary human geography that without an understanding of the techniques employed an appreciation of the discipline's content is impossible; quantitative analysis is increasingly important in many areas of society, and a well-educated citizen should be able to evaluate the presentation of numerical material; and, following on from the last point, numeracy is a skill that employers increasingly demand.

Such arguments, though valid, are self-sustaining and, to a considerable extent, conservative. If geography can only be understood by those with a knowledge of quantitative methods, then the type of geography produced in that way is likely to be sustained to the detriment of others. (The same is true in more detail too, as Bennett, 1985, argues in his case for Bayesian and against Neyman-Pearson statistics.) Further, if a substantial number of academic human geographers have followed this route themselves, and are extending it in their own researches, then the definition of disciplinary progress is very much tied up with increased technical sophistication. This increases the volume of material that students have to appreciate, which means even more technical training. (In general, students are unwilling/unable to tackle this, and as a result most departments do not press them too hard – the consequence for student recruitment is too unpalatable. Thus there is a growing gulf between the small group of researchers advancing the technical sophistication of their work, the bulk of their colleagues who are 'left behind' – many sticking with, and frequently misusing, the simpler procedures – and the vast majority of students who hardly start up the road.) Numeracy is necessary; but not too necessary because the the sovereign consumer may protest.

Literacy skills are developed in all social sciences and quantitative skills in most – in several (economics and psychology in particular) to a greater extent than in human geography, but again with problems of consumer resistance. A third set of skills developed mainly for geographers alone is graphicacy: the ability to present and interpret information in graphical and, especially, cartographical form. After some years of neglect, these are becoming more prominent in degree offerings, to a considerable extent aided by technological developments in data acquisition (notably remote sensing), collation (geographical information systems), analysis (spatial statistics), and presentation (automated cartography). These must be supplemented with direct observation and the acquisition of particular data-collection skills, however, and the fieldwork component is another particular feature of geography degrees.

Within many degree programmes, some of these skills – notably those of numeracy and graphicacy – are taught separately from the substantive courses. For numeracy it is argued, with a great deal of validity, that this is necessary because quantitative procedures must be taught as a linear sequence; one needs the basic groundwork prior to the applicable routines, and prior to an appreciation of the techniques used in the cited research papers. And then, once the groundwork has been laid, students can use the techniques in an independent study (a project, dissertation, or thesis), in which they combine a range of skills: problem definition, data collection, analysis, and report.

The skills taught are very much oriented towards work in the empiricist/positivist tradition, and most student independent studies have a positivist flavour to them. This, indeed, is the implicit goal of most degree programmes, for many of the substantive courses – a majority of which are systematic, with titles such as industrial geography, historical population geography etc. – have a similar orientation and, as the quoted advertisements show, a major aim is to turn out graduates with basic skills in empirical description and pragmatic problem-solving. To some this aim must be expanded, with greater attention to the provision of skills that graduates can apply. (This has been argued especially by some physical geographers, e.g. Worsley, 1979, 1985, who believe that the breadth of geographical education, at school and university, prohibits full development of the needed skills for a properly-trained graduate. They have a much clearer vocational goal. Human geographers are less vocal in this respect. However, many of them are strong in the defence of separate geography degrees and against the broader base in the social sciences being provided for human geographers. They promote the fragmentation of knowledge at both the inter- and intra-disciplinary scale, thereby preferring the chaotic conception of 'geography, only geography' to the more rational abstraction of 'geography in social science'.)

Although departments vary in the relative stress they place on various skills-training courses, all place them at the centre of their degree programmes: they are as basic as the introductory courses provided (usually one each in physical and human geography). In their substantive courses they are, with a few exceptions, much less prescriptive, largely as a consequence of appointment policies. As noted above, most departments have appointed staff in order to provide a breadth of topical coverage, and have then allowed the individuals concerned to develop their own substantive courses. Apart from some attempt to provide basic courses in generally popular subjects, such as urban geography, in general what is taught in the substantive courses reflects what the staff want to teach, which in most cases is their research interest; there is little attempt to provide a coherent degree structure – perhaps because it is impossible?

Most degree programmes are set in the empiricist/positivist mould, therefore, preparing students to be problem-solvers and reflecting the fragmentation that is typical of the systematic fix. Alongside this focus on technical control, via skills and their application, the other goals of geography – mutual understanding and emancipation – receive relatively little attention. A degree in human geography is increasingly a general training rather than a general education. Many students receive virtually no exposure to the humanistic and structuralist philosophies: they are denied access to what one might term the regional cultural geography that would be part of either a humanist or a realist programme. Their experience of the world is that of the outsider; their insights into life in places – even home or next door, let alone the rest of the world – are few. Thus it is not the concentration on skills *per se* that is the problem, but the context – the systematic fix and the spatial (distance only) fix – in which they are taught; the applications are biased towards particular types and uses of geography.

Two rationales can be offered for this 'bias' in the structure of degree programmes in human geography. The first is the consumer-sovereignty argument. Empiricist/positivist spatial science rose to dominance in the 1960s, and has remained there, because this was what society in general wanted; it has brought status to the discipline and enabled it to prepare graduates for useful careers, which is what they want. (Interestingly, and much to the chagrin of some spatial scientists, many students opt, where the choice is available, for courses that promote awareness of places – not traditional regional geography, but still largely empiricist – rather than technical systematics.) From this position of dominance (one which is increasing with the retirements – early in an increasing number of cases – of people socialized into different views of geography and with a lack of recruitment to replace them), the spatial scientists are able to manipulate the reproduction of the discipline. Academic freedom allows those in post to reorient their work, but relatively few do. Demography is on the side of continued dominance by the empiricist vision.

Secondly, and linked to the first, there is the argument that human geographers cannot change society and therefore they must prepare people to survive in it; to do otherwise is to lose credibility. The sort of training programmes just described at least give graduates the foundation for a career. What would a programme built around greater mutual awareness provide; a better quality of life? But in late capitalist society, quality of life is firmly linked to economics, so what use is a 'good liberal education' if you can't get a job? And as for emancipation, this provides a foundation for nothing in the short term (as well as being, to some, potentially subversive). Its goal is to demystify the mechanisms. But what do people do when they are demystified, when they understand why they are unemployed?

The pragmatism of these arguments is hard to gainsay, for people do need to be able to live after they graduate, and their ability to advance themselves is likely to be enhanced if they have been provided with transferable skills. (Although, as noted above, potential employers are increasingly unconcerned with the substantive context in which those skills have been learned, when there are fewer jobs than graduates it is always safe to have followed a programme in which the substance is potentially useful too.) This is a depressing prospect, for it provides a major straitjacket to education and the development of applied science. As argued in chapter 6, the case for applied geography as the promotion of awareness is a powerful one. Given what we have already done to the earth, what we are now doing, and what we have the potential to do, the need to appreciate ourselves, each other, and our environment is pressing. The case for emancipation is equally strong, for people should be liberated not caged, should be enabled to subject their society to criticism and to work for its improvement, not confined to a particular ideology of development and progress. Therein lies the nub of the problem. Mutual awareness is acceptable, because it is non-threatening; it is not useful though. Emancipation is threatening, because it seeks to unmask the ideology. Capitalism is robust enough to withstand many such threats from ivory-tower academics – because sooner or later their disciples must live in the 'real world' – and so can tolerate a few radical social scientists, but no more. (Recall Harries's, 1976, response to radical criticisms of his work in the geography of crime that 'a revolutionary polemic usually guarantees the American academic a place in the unemployment line' – p. 100 – and 'I would be the first to admit that my experience in a land-grant university may have fostered a greater awareness of pragmatism than might have been the case had I located at an institution without an explicit public-service role' – p. 101.)

Clearly human geography occupies an intermediate position between the professions – e.g. medicine, law, architecture, engineering – whose degree programmes are carefully organized to meet the requirements of outside bodies, and the humanities (plus some social sciences) which are clearly providing an education with very little in the way of training (other than in literacy). The general trend in human geography is towards the former, however, though as yet without the institutionalization. (When it still had a substantial number of postgraduate training awards available, the Human Geography Committee of the Social Science Research Council did conduct an accreditation exercise, which most departments passed.) There has been little argument for the provision of structured training programmes and outside certification (though some, e.g. Clayton, 1985, argue for a common core of material in all degree courses), but the overall drift is for the provision of transferable skills that can be used in future careers. This is at the expense of education. Human geographers have isolated themselves and their students in their own degree schools, restricted access to the wider strands of thought and debate in the social sciences, and focused their work along empiricist/positivist lines. Most would claim that they are not being prescriptive, and yet in aggregate they are. Education is being sacrificed for a particular form of training.

Further and higher education institutions are centres for research as well as teaching; indeed for disciplines such as human geography they are the only places where research is carried out, except in a small number of research institutions whose work is closely tied to the needs of clients and sponsors. It is part of the ethos of such institutions that there is a symbiosis of research and teaching, and that it is the contact with people on the research frontier which gives the teaching its particular qualities. In some areas of empiricist/positivist human geography this is no longer the case, however, for the research frontier is way beyond the final-year undergraduate in its technical sophistication – which is seen by some as providing the rationale either for more undergraduate courses or for postgraduate training in order to produce problem-solvers with the needed expertise.

This widening gulf between the research frontier and the teaching task is only to be expected. Career opportunities for academics are determined largely by their research performance, and so there are strong demands for them to move forwards in acceptable directions. The lead time needed for students to catch up with them must become longer: more skills must be learned; more textbooks assimilated etc. This is recognized in other disciplines by the introduction of longer degree courses and by the growing numbers who take postgraduate degrees. But this means that training dominates. Is this the way in which human geography should go?

Several alternatives are canvassed. More of the introductory material could be taught in the schools, perhaps, but this makes the school a place of training rather than education. Departments could become more specialized, with research teams focused on particular subject areas, with students able to transfer once they are clear what they want from their final year(s) as undergraduates. But in the end, they all point to the same conclusion: human geography in higher education is biased towards training rather than education, and that bias seems to be growing. There is a need for a new balance, one that allows – though does not prescribe – less emphasis on techniques and on fragmented systematic knowledge and more on the promotion of awareness, at both the actual and the real level. Education should draw people out, not constrain them.

HUMAN GEOGRAPHY IN SCHOOLS

The nature of human geography in the schools is closely linked to what is being done in the universities etc., though usually with a lag of a few years. In Britain, this has been because school curricula from the age of 14 on have been oriented to examinations (O- and A-level) set by outside bodies dominated until recently by the universities. A major function of those examinations has been to assess pupils' potential for further and higher education, and so the subject syllabuses have been devised for that purpose, by the people who will be teaching the successful pupils.

Because of this, changes in university orientations have been reflected in the school examination syllabuses. For geography in general, and human geography in particular, this was the case in the 1970s when the empiricist/positivist approaches replaced the empiricist/regional approaches, especially at A-level. What are known to some as 'concept-based' syllabuses replaced those that were 'location-based', and the implicit universalism of positivism (and models such as those of von Thunen and Weber, Burgess and Christaller) was introduced. Places no longer mattered, except insofar as they had physical features which might 'distort' the models; they were merely the testing grounds. Thus school geography, like university geography, has been linked to the notion of training in the solution of empirical problems.

More recently, these developments have been challenged, largely by those who teach in the schools. (There were always some reservations about the changes and the demise of regional geography, but largely on the grounds that the 'new geography' was hard to understand and teach.) That challenge is founded upon a belief in the role of human geography as a necessary part of an *educational* experience for all, not as a training programme for a few. (Some are not sure, of course: see S. Gregory, 1978.)

The outcome of these challenges can be seen in recent developments. With regard to syllabuses to be used by examination boards for pupils in the 16+ age group, for example, the introduction to the *National Criteria for Geography*, which all syllabuses must apply, states that:

> Geography is concerned to promote an understanding of the nature of the earth's surface and, more particularly, the character of places, the complex nature of people's relationships and interactions with their environment and the importance in human affairs of location and the spatial organisation of human activities.

> Geographical education may be seen in terms of knowledge and understanding, skills and values. The term 'values' is included to indicate that important topics in geography syllabuses have obvious social and political dimensions and cannot properly be understood without taking account of the attitudes and values of those involved.

Similarly, the *Core Syllabus for A-Level Geography* identified seven points in answer to the question 'What ought a student to have gained as a result of taking an A-level course in Geography?':

1 An awareness of certain important ideas in three areas: in physical geography; in human geography; in the interface between physical and human geography.
2 An appreciation of the processes of regional differentiation.
3 Knowledge derived from a study of a balanced selection of regions and environments, linked with a broad understanding of the complexity and variety of the world in which the student will become a citizen.
4 An understanding of the use of a variety of techniques and the ability to apply these appropriately.

5　A range of skills and experience through involvement in a variety of learning activities both within and outside the classroom.

6　An awareness of the contribution that geography can make to an understanding of contemporary issues and problems concerning people and the environment.

7　A heightened ability to respond to and make judgements about certain aesthetic and moral matters relating to space and place.

Do these documents not show a better balance of education and training than is typical of higher education?

Statement of such goals is insufficient, however; it is their implementation that is crucial, and how this will be done is not clear. Certainly existing syllabuses do not substantially promote the awareness – let alone the emancipation – that might be anticipated. In general, the cultural variety of the world is ignored, except insofar as certain cultural stereotypes (often implicitly, if not explicitly, racist and sexist) are used to 'explain' why some people do not act as 'rationally' as they would if they understood Weber, Christaller et al. But whereas cultural variety is relegated to an inferior position, environmental variety is not. In the late 1970s, for example, the Northern Universities Joint Matriculation Board introduced a new syllabus C, which aims to:

> educate students so that they can arrive at an understanding of regional systems at a variety of scales of analysis ... against a background of:
> (i)　an understanding of the effects of physical constraints and stimuli on development;
> (ii)　an understanding of the historical perspectives relevant to the present situation;
> (iii)　an appreciation of the influence of institutional factors in shaping the growth and structures of regions;
> (iv)　a training in the basic skills needed for the application of these ideas in terms of geographical enquiries and investigations.

Within the syllabus, the themes of urban and economic development were to be studied 'against a background of: the environmental setting, the historical perspective, the effects of institutional factors' in either the EEC or North America; and the themes of agricultural development, settlement pattern and economic growth were to be studied in one region of the humid tropics and arid lands under the direction that:

> *The characteristics of the environment* should be studied as a background to the themes and as relevant to their full understanding.

All of this suggests a regional geography that promotes awareness of cultural variability. But in its interpretation, the physical environment is taught and examined as 'classical' physical geography, whereas there is no separate teaching (let alone examining) of the history and institutions (i.e. culture). The environmental fix maintains a strong hold. So does the spatial fix, for 'location theory, network analysis, quantitative methods, the use of models and so forth should be used as appropriate. . .' And the systematic fix, for parts are stressed rather than wholes. Furthermore, the choice of regions to study excluded 'difficult' areas, such as China and the USSR, which do not necessarily fit the definitions of development.

The basic problem underlying issues such as this is the historical constitution of the discipline of geography, with the continued strength of the environmental fix and the recent dominance of the spatial and systematic fixes. Those who teach geography in the schools are unprepared by their experiences in the universities for a discipline with links that are as strong with the social sciences as they are with the physical sciences. Thus the intentions of the JMB syllabus cannot be fulfilled because people lack a realist understanding of society, and the best that they can offer is a voyeuristic awareness of 'other sorts of people' in other places. Physical geographers like Gregory and Worsley may feel that their discipline is badly served because students are inadequately prepared in the sciences. At least they could be, for the problem is that pupils doing geography do not select scientific courses too, not that they could not. For human geographers the position is much worse, since the general social science background is not available. Economics is now a major subject in the schools, but it is neo-classical economics which is as empiricist as the industrial geography based on it. The whole structure of British education is such that a full appreciation of societies is not available, except through the study of history, which is rarely social scientific in orientation.

Although there are blueprints for change, therefore, it is doubtful whether the current constitution of geography can meet them. If we interpret the national criteria and the core syllabus as presenting a major case for human geography as the promotion of mutual awareness, then is the orientation of the subject correct? Is the link with physical geography too strong? Are we offering a holistic view which studies societies in their environmental contexts rather than studying environments and then putting people in them? If the latter, then are we not in great danger of promoting environmental determinism and cultural imperialism?

The key problem is the degree to which we are prepared to go beneath the surface appearances, to plumb the empirical and the actual to reveal the real. Most of the 'new' regional syllabuses, such as that described above, do not, because they study regions in isolation. It is accepted that regions differ, but it is assumed that the reasons for that are entirely internal to those regions. They are not placed in their global context, in a world-economy which respects no political borders. To study places in isolation is to tear them out of context, and it leads to what Peter Taylor (1986b) calls the 'myth of developmentalism', the belief that change can be generated internally and independently of any wider context. The regional geographies being promoted do not take account of his important identification of three scales – experience, ideology, and reality (p. 67); almost all of the attention goes to the first, and virtually none to the last.

One of the problems of promoting realistic understanding via school syllabuses is clearly that it is perceived as subversion, because it promotes a critical stance towards capitalism and its ideology. Sir Keith Joseph's (1985) presentation of desirable values in geographical education has already been noted (p. 119), and elsewhere, in a Green Paper on *The Development of Higher Education into the 1990s*, the Secretaries of State for Education and Science, for Scotland, for Wales, and for Northern Ireland warned those in higher education:

> to be concerned with attitudes to the world outside higher education, and in particular to industry and commerce, and to beware of 'anti-business' snobbery. The entrepreneurial spirit is essential for the maintenance and improvement of employment, prosperity and public services . . . higher education needs to foster positive attitudes to work.

Thus while the efforts of groups seeking to unmask the racial and other prejudices in syllabuses (e.g. Gill, 1982) may be provocative (and, to some, painful) they are no more than that. (They are, of course, extremely valuable in at least drawing attention to the Eurocentric paternalism of much geography.) Others are, however. Huckle (1983), for example, has proposed a socialist approach to school geography in which:

> lessons can not only sustain prevailing beliefs and attitudes but may also allow pupils and teachers to examine alternatives openly and critically. There is scope for committed work in schools and the current crisis of capitalism provides geography teachers with significant opportunities. They teach about industry, towns, developing countries, resources, pollution, and world trade and are therefore well placed to explore the mounting contradictions of a mode of economic

and social organization based on inequality and exploitation. By dealing with unemployment, inner city decay, underdevelopment, resource depletion, environmental deterioration and the global economic order in a manner which acknowledges conflict and seeks political literacy . . . requires politically aware geography teachers who recognize their role as agents of social change. (p. 151)

And the Association for Curriculum Development in Geography has launched the journal *Contemporary Issues in Geography and Education*, including objectives:

to examine the ideological content of geographical education in relation to its political context,
to demonstrate the relevance and importance of humanist and radical ideas for teaching and research in geography,

and:

to encourage the realisation of the links between critical understanding and the active transformation of the world in which we live.

These, very clearly, are set in the 'applied geography as emancipation' mould. They promote geography as the study of the insider's as well as the outsider's world, *and* as the study of how those worlds are created.

To many, such manifestoes are the work of radicals – but not dangerous radicals, just annoying ones, because they are unlikely to have much impact on the ruling ideology. But their thrust is hard to reject. In school education today geography is unbalanced: just as in higher education, it is empiricist, emphasizing training for problem-solving and, despite many good intentions, avoiding understanding. Within human geography it underplays mutual understanding and the important role of place in the constitution of society. More importantly, it fails to explore what it is that produces geography, other than variation in the physical environment. By not seeking to understand societies, only describe them, however sophisticatedly, it fails to educate people, fails to help them see what it is that governs their lives. It is part of the ideology of capitalism, promoting the ruling ideas of that mode of production.

THE ORGANIZATION OF AN ACADEMIC DISCIPLINE

As noted earlier in this chapter, academic disciplines are basically collections of individuals most of whom – the great majority in human geography and like disciplines – teach in a collective context (a department) but conduct their research individually. In addition to their teaching and research, both required of them by contract, most are involved in academic administration at a variety of levels, within their department and institution and in a range of related bodies (research councils, learned societies etc.).

The prospects for individuals and small groups in mass societies are relatively slim, unless they have access to powerful positions within economy and polity. Thus in a variety of ways they need to act collectively in order to protect and promote themselves and what they do. One way is through trade unionism, which promotes the interests of all undertaking a particular role (assuming that the union is able to win bargaining rights), such as teaching in a school or university. For academics this activity covers contractual issues (salary levels, terms of service etc.) and promotes the elusive concept of 'academic freedom', but is not involved with the detailed nature of the work done nor with the promotion of particular disciplines; a union can promote all academic activity but not some aspects of it to the detriment of others. (Academic unions are notoriously weak and poorly supported. In Britain, the Association of University Teachers negotiates with Vice-Chancellors over pay and conditions; but many Vice-Chancellors are AUT members.)

Each discipline needs its own equivalent of a trade union, therefore, to promote its members' interests. For most disciplines this involves (a) promoting the educational role, to ensure a guaranteed place and size within school and higher education curricula, and (b) promoting the research role. Until very recently, in Britain the two have been treated as one within the university system. Now there are attempts to decouple them, to allocate resources for research separately from those for teaching: the latter provide baseline funding for all and the former are allocated according to proven excellence and 'relevance'. As yet, separate 'unions' for the two roles have not been proposed.

Within academia in general there are two types of 'trade union' for individual disciplines (or groups of disciplines; or subdisciplines). The first is the *professional association*, which controls both the teaching activities – by defining what should be in the syllabuses and by validating degree and other programmes – and the professional activities of the individual practitioners, academics and others. Medicine and law are the best examples of these, with their professional associations (e.g. The British Medical Association and The Law Society) operating closed shops: they are recognized in law as the validating bodies, so that only those who meet their criteria are allowed to practise, that is, to call themselves doctors; and they set standards of professional conduct which members must adhere to, and which if broken can lead to the miscreant being 'struck off' and banned from practice. The last few decades have seen many attempts to extend the range of professional associations, to create more closed shops and so gain monopolies over the provision of certain services – for example, accountancy.

The second type comprises the *learned society*, which is a voluntary body open to all who wish to join and who meet fairly minimal criteria. Fellowship of the Royal Geographical Society is open, for example, to:

> all those who wish to promote its object, defined in its Royal Charter as 'The Advancement of Geographical Science'.

The Institute of British Geographers has a similar broad brief, stating to would-be members that it:

> exists to further the discussion and advancement of geography, and to promote the interests of professional geography and geographers on a national and international basis through its publications, its Annual Conference, its active promotion of Study Groups devoted to specific issues and topics in the discipline, and its encouragement of international symposia.

For these societies, promotion activities are less easy than in the case of professional associations because they cannot claim to speak for all practitioners – no body can claim to speak for all geographers in Britain, for example, in part because it is very difficult to decide who 'all geographers' are, though it can claim to be representative of all geographers employed in certain occupations. Such learned societies have two major roles: (1) to act as pressure groups; and (2) to provide collective services. As pressure groups, they make representations on behalf of the discipline; their success depends very much on the qualities of the individuals making the representations. As providers of collective services their main roles are in the organization of conferences and other meetings and in the publication of journals, monographs and other materials.

In Britain, there are three learned societies which represent geographers nationally, plus several regional ones. The longest-established is the Royal Geographical Society, which in the past has promoted a particular type of geography (exploration, scientific as well as voyeuristic) but which now, from its London base and its contacts in the House of Lords and elsewhere, acts generally in the interests of geographers – of whom practising academics are a relatively small minority. The next oldest is the Geographical Association, concerned almost exclusively with geography in education and an increasingly successful pressure group for geography at schools. Finally, there is the Institute of British Geographers, much the smallest and – despite its wide membership criteria – almost exclusively focused on the interests of academic geographers. It was founded in the 1930s to provide an institutionalized publication outlet (because the RGS was perceived to have negative attitudes to certain types of geography, mainly human) and to organize conferences. These have remained its major functions, though in recent years pressure-group activity has increased, some of it in concert with the RGS and the GA.

There are other bodies, such as the Conference of Heads of Departments of Geography in Universities and the Conference of Heads of Departments of Geography in Polytechnics and Colleges, each of which meets regularly, if not frequently, and occasionally makes representations on behalf of geography teachers in higher and further education. (They do not promote research into educational practice, however, a field only thinly populated – despite the excellent *Journal of Geography in Higher Education*.) But those representations must take a particular form, since they must be acceptable to all members: thus in framing a response to questions from the UGC in 1984, the officers of the University Conference received strong negative reactions to draft comments which implied that some small departments may not be viable. They could not claim to speak 'for university geography', but only for a group of vested interests. Such a situation will undoubtedly frustrate a proposal of Brunsden's (1985) to create a National Coordinating Committee for Geography. He notes that each of the societies and conferences has been sensibly active, indeed 'assiduous in ensuring that matters of national importance are considered' (p. 464), but notes that the fragmentation makes the coordination of responses cumbersome, 'so that the impact of a whole profession speaking with one voice is reduced'. However, according to my argument here, geography is not a profession and, except in the most general terms, one voice to represent (or to speak without checking whether it is representative of) geography is virtually impossible. Brunsden's National Committee would need the following brief:

– to identify problems of national concern to geographers;
– to approach societies (and to receive advice from societies) on issues of concern;
– to identify the correct body and/or individuals to respond to these problems;
– to approach the Education Committee of the Royal Society and the

British Academy on these issues and request their help in advising government;
- to ensure that a response is made within the timescales relevant to the problem.

It, too, sounds cumbersome and time-consuming. More importantly, it assumes a professional consensus which is not, will not, and should not be there. Geographers should and will respond, individually and collectively as appropriate. But 'geography' cannot respond.

Because it is possible to be an active academic geographer without belonging to any of these learned societies – to be a freerider – what they do, and how they present geography to outsiders, depends very much on the interests of those who chose to join and to be active. The context will be a strong influence on what people want from a society, however. Within the IBG, for example, the current problems with regard to the allocation of resources to higher education and research in general and to geography in particular lead to demands for active promotion of the discipline, as a 'relevant' discipline (i.e. one that trains people to solve problems and undertakes useful, control-oriented, research). In the 1960s and 1970s, when expansion was taking place, the main focus was on the internal activities – more meetings, especially specialized ones (organized by the specialist Study Groups), and more publications (a greater frequency for the main journal, *Transactions*, and the launching of a new one, *Area*).

There is a difficulty for learned societies such as the IBG in performing these two roles. On the one hand, they have to undertake external posturing, promoting the discipline as a whole and claiming to speak on behalf of all, while on the other there is a need to ensure the academic freedom of the individuals. For the former, it is necessary to present a united front, to promote geography as a whole, but if the attitudes developed for this are carried over into the second role it carries the risk of the creation of what is termed the *guardian ethic* (Elliott, 1974), a variant of 'the cult of the expert'. For outsiders, it is necessary that the officers and other elected representatives of a learned society promote their members as the best (if not the only) people within society qualified to advise on certain subjects and to undertake certain tasks. But it should not be acceptable that those same elected officers and representatives seek to govern the activities of the members themselves, deciding what is and what is not acceptable for discussion at a conference, what groups can and cannot come together, and so on. This is to create a monolithic professional association out of a learned society; it implies paternalism and censorship, and is likely to be alienating and discourage membership.

To outsiders, therefore, a learned society must present a united front, in order to speak for geography. For its members, however, it must *promote anarchy*. This is not a synonym for chaos (except in the vernacular usage!) but a description of a particular form of social organization in which individuals are free to associate in communities, to shift around, to dissolve and reform them etc., with the minimum of central control. (The desirable minimum is zero — but financial aid is always welcome.) Maximum flexibility is the goal. It is messy but by far the best way of structuring academic life: it creates problems when anarchy within has to be promoted beyond as united, but it is necessary. The guardian ethic is inimical to a discipline such as human geography, because it involves attempts to professionalize what should not be professionalized. A realist social science must be based on the maximum freedom of thought, expression and association.

Freedom of expression is a difficult issue for learned societies, both oral expression — the right to speak at a meeting — and especially written expression — the right to publish in its journals etc. Most societies publish journals covering the entire discipline, such as the *Transactions* and *Area* of the IBG, *Geographical Journal* of the RGS, and *Geography* and *Teaching Geography* of the GA. Within the IBG, *Transactions* is relatively unpopular with the members (see Clayton and O'Riordan, 1977). Because of its breadth few find many of the papers of interest and prefer more specialist journals, almost all of which are produced by commercial publishers. Journals such as *Transactions* represent the inertia of the integration fix — but specialist journals pander to the systematic fix.

For geographers today there is a wide range of journals produced and the problem discussed in the previous paragraph is not seen as serious by most. Much more importantly, there is the problem which faces all journals with regard to what is, in effect, 'academic censorship'. In almost all cases, individuals have no right of publication, even in the journals of the societies to which they belong. (The major exception is that most publish the addresses of their elected presidents 'uncensored'.) To get a piece published you must pass a series of gatekeepers, the journal editors and the 'expert' referees that they consult. In most cases decisions are made on: (1) the novelty of the piece — is it saying something new?; (2) its quality — is what is said the result of good research?; and (3) its presentation — is the material presented effectively and efficiently? The biggest problem is with the second of these, for what is good research? In some technical areas absolute criteria are available — or are they, for there is considerable debate over many statistical procedures? But in general it is a subjective judgement.

Subjective judgements can amount to censorship. Very often this will be without malice; the referee makes an honest attempt to provide an 'unbiased' report, but it is nevertheless a restriction on freedom of expression. But most academics have anecdotal stories of explicit censorship, too, of papers being dismissed because of their approach. (In periods of major shifts within a discipline, this can lead to wholesale 'suppression' of certain points of view. The consequence is usually a new journal, such as *Geographical Analysis* and *Antipode*.)

Although explicit censorship may be rare, there are plenty of recent examples of worrying events, such as authors having papers rejected because they are of interest to only a minority of the members of a learned society. The gatekeeping role is a difficult one, but if not operated as liberally as possible it is equivalent to imposing an orthodoxy – which in the long term is likely to be self-defeating. But to be too liberal may be self-defeating too, because it will undermine the commercial viability of a journal, whether produced by commercial publisher or learned society. The result then may be the 'nonconventional' or 'underground' publication – the circulation of papers by individuals to others; this is not censorship, but it may mean that certain materials are much more difficult to access than others.

Similar gatekeeping takes place with regard to the publication of books. Academics are involved in the production of the following six types.

1 The instructional text, providing introductions to needed skills at particular levels; for human geographers, statistics texts are the main example of this type.
2 The synthesis text, which summarizes knowledge in a particular field (e.g. the whole of human geography) or, more usually, some part of it (e.g. urban geography).
3 The reference book, such as an atlas or dictionary.
4 The research monograph, which reports in full detail on the results of a programme of research.
5 The essay, which aims to stimulate interest in a topic.
6 The reader, a collection of new or reprinted essays either intended as an alternative to the synthesis text or providing a collection of brief research monographs or essays. (A subtype is the volume of conference papers, loosely linked around a theme and close to the collection of research monographs.)

Of these six, the first three are the most popular with publishers because they have the largest potential markets – especially if they are written for the large, introductory undergraduate classes. (American publishers will invest a great deal in the production and promotion of such a book, and are not very interested, in general, in those aimed at smaller, more tentative markets. Some, it seems, would rather have a competitor for an established book in a field than a new book seeking to develop another field. British publishers are more liberal, and are much more prepared to back small-market books – places matter?) Types 4 and 5 are the more risky, especially if they are by unknown authors, though publishers do seek to promote quality, even in small markets (Ashby, 1981), and some are prepared to back 'promising' authors.

Once again, the strategy is – necessarily, in the context of a capitalist system – conservative. It is supportive of the status quo and promotes conformity rather than freedom. It is valid for a professional discipline, but is it always in the best interests of a realist social science?

As to freedom of oral expression, most societies run more or less open conferences with only limited constraints on, for example, time. (The Association of American Geographers allows all who apply to present *one* paper at its annual meetings, but in most cases little more than 20 minutes for the presentation.) To some this lack of any censorship (or, in the politer language, of refereeing of proposed contributions) is detrimental to the quality of the conference – though it is not always clear whether they are criticizing the mode of presentation (which often leaves much to be desired) or the material presented. Clearly, 'open house' presents organizational problems – too many complaints about overlapping concurrent sessions; too many small audiences – but organizational complexity (even chaos?) is surely preferable to subjective censorship.

There are, of course, other gatekeepers who play important roles in the direction of a discipline's development, such as those with influence on the allocation of research grants and studentships and on the making of permanent academic appointments. It is hard to assess their importance, just as it is difficult to evaluate how influential journal editors are. Certainly, the discipline of human geography in Britain is too large for the views (whether dogmatically conservative or agnostically liberal) of a few to dominate, unlike the situation in the past when there were many fewer geographers (a situation still present in some countries: see Lichtenberger, 1984, on the 'popes' – misprinted as 'papers' – of German geography, and Bosque-Sendra et al. 1983, on the centrally determined university curriculum in Spain). But some individuals are clearly more influential than others, and their views, though not decisive, could well be steering the discipline in particular directions.

IN SUMMARY

The general tone of this chapter is pessimistic. Having outlined an approach to human geography that is meant to be liberating, I find that the institutional context is at best neutral towards the goal, at worst working against it. As so frequently is the case, this does not necessarily imply either strong opposition to the approach or a wish to censor it, but the 'pragmatic realities' of life in a materialist society provide the censors. This is not for one moment to suggest that the struggle should be abandoned, but it is clear recognition that it will be long and hard.

9

Moving Forward

It is pointless modifying conventional geography, arming it with
additional mathematical or behavioural tools, linking geography
with a vague moral humanism, or conveniently losing ourselves in
the immediate intricacies of the human experience.

Richard Peet, 1985, p. 10

Where next? Having got this far and being unprepared, unlike Richard Peet,
to overthrow all of the geographical baggage that I have accumulated en
route, how should I practise what I preach? What geography should I be
doing? To answer this, I tackle three related topics: the links between
empirical and theoretical research; the issue of autonomy; and the questions
of practice. In this way, I hope to tie together some of the issues raised in this
book and to answer a critic who has identified my 'mid-career crisis' and my
'self-confessed, frenetic and frequently undisciplined English empiricism'
(Mercer, 1985b, p. 147). I trust that already in this book I have already
countered the caricature of my position that:

> the data produced by Thatcher's 'liberal democratic state' are there to
> be taken at face value, to be analysed speedily through the medium of
> packaged programmes and then distributed for mass technocratic
> consumption. If the data are not there, there is no research.
> Presumably, then, it follows that if vital concepts such as 'ideology',
> 'power', 'class' or 'hegemony' cannot be reduced to simple quantita-
> tive formulae they are deemed not to exist. It is as simple as that.
> (p. 147)

Nevertheless there are many loose ends, and much to debate, for I do *not*
claim to be right and to have all the answers. I recognize that many
problems remain, that my work is evolving inconsistently; all I am doing is
sharing my thoughts on the way forward – which is certainly not, to me,
empiricist, quantitative or otherwise.

THEORETICAL AND EMPIRICAL

Faced with arguments such as those presented here, many people have difficulties in identifying links between what may be termed theoretical and empirical research. In particular, many perceive problems in justifying empirical work, and provide schemes that suggest a series of 'levels' at which work may be conducted.

Typical of such approaches is an essay by Herbert (1979) which identified four levels of analysis related to urban problems: the first two are referred to as structural levels.

1 The political-economy level, where the main thrust is 'to expose the basic structure of the social formation' in a 'rather exacting mental discipline' (p. 4).
2 The resource allocation and provision level, concerned with 'ways in which the economy or social formation begins the process of allocating power and resources' (p. 5).
3 The urban managers, 'who hold power at an intermediary position in the allocation system and whose decisions directly affect the urban environment' (p. 8).
4 The local environment, which is 'concerned with the socio-spatial outcomes of structural processes' (p. 6), through areal, ecological, behavioural and prescriptive analysis.

These four present a series of 'broad, inter-connected fronts along which geographers are pursuing analyses' (p. 8). All four 'have a role in the future form of our perspective' (p. 3), but there is no evidence of how they can be integrated to provide a holistic view. Theoretical understanding and empirical research are separate entities, it seems.

More recently, Phillips and Williams (1984) have argued somewhat similarly. They accept that rural deprivation, the focus of their work, is:

inextricably linked with the whole issue of urban deprivation . . . Both are the products of the same forces that emerged from the dynamics of late industrial capitalism – structural shifts in the economy and changing demand for, and role of, labour (p. 236)

which leads to the conclusion that:

a political economy approach is essential to furthering the understanding of rural social geography (p. 237) with such an approach defined as the analysis of the 'production' of constraints (p. 15). In summarizing their own work, however, they conclude that:

> the theoretical understanding of political economy as it affects rural areas is inadequate . . . [so] many of the substantive chapters are based upon managerialist and behavioural studies of social processes, with the implicit (rather than explicit) notion that these are located within the larger social formation. (p. 237)

This almost apologetic rationale for separating the results of empirical work – analyses of the worlds of events and experiences (p. 51) – fails to identify the symbiosis of empirical and theoretical work. (It also assumes that the concept 'rural' is a rational abstraction.)

To reiterate again my view of the structure of research in human geography, I will build this argument on three foundations:

1   A theory of society is needed that can tell me *why* people act as they do, that can identify the imperatives for action;
2   That theory will tell me *why* people do as they do, but cannot predict *how* they will act, because all action takes place in the context of 'contingently related conditions' (figure 4.3) which mediate the actors' interpretations of the imperatives; and
3   The 'contingently-related conditions' are historically and spatially specific, are not themselves predictable, and therefore cannot be used to predict further conditions and actions.

In other words, I distinguish between *how questions* and *why questions*, between those concerned with specific realizations and those concerned with the general forces.

There is a third category: *what questions*. These focus only on the empirical worlds, describing them through the eyes of either an observer or a participant. Answers may be the preface to research, with questions relating to how and why being posed (as suggested by figure 5.1), but no more. The vast majority of such studies will be either empiricist or idealist, collections of 'facts', akin to stamp-collecting, except that the facts in that activity are virtually incontrovertible. The 'facts' of empiricist work are not. They, like all other facts, are theory-laden, for one cannot observe, record and describe without a theoretical language, a set of membership laws. Peter Taylor (1983) has argued not only that combining theories from different epistemologies (realist, positivist etc.) is impossible because they are incompatible but also that non-realists (implicit supporters of the status quo), 'are ultimately concerned with *order*. It is for this reason that they do not need a theory of the state: the order imposed by the current state system is accepted as given' (p. 13). In other words, those who want to answer only *what* and *how* questions do not need a theory that offers answers to *why* questions; they only need that if they are radicals, wanting to alter the situation. He accepts the value of empirical, 'what and how' research, however: 'The empirical tradition of the status quo group, for all its faults, provides, and will continue to provide, a wealth of information for critical theorists to evaluate' (p. 16). But full appreciation of that information requires an appreciation of its theoretical base: status quo empiricists may not need a theory (of the state in this example), but they have one. The goal must be to replace it by the realist, for, as Hudson (1983) notes: 'It *is* correct to identify different levels of analysis but equally incorrect to suggest that these have to be dealt with separately within different theoretical frameworks, and then welded together' (p. 33). What, how and why questions should be part of a holistic search for understanding.

Empirical research is concerned with how and what questions, and theoretical research with why questions, therefore; the former seeks to understand action in the context of interpretations (figure 4.2), whereas the latter seeks to understand the reality of the base and superstructure that are being interpreted. The two are interdependent. Empirical researchers need a theory of the real within which to set their investigations, even if they are not contributing to the development of that theory; the questions that they ask are theory-led, even though probably stimulated by empirical observations, and the theory ensures that the topics that they ask questions about are rational abstractions, not chaotic conceptions. Their how questions are set within a system of answers to why questions – a system that the answers to 'how?' will enhance. This does not necessarily mean that all reports of empirical research must be introduced by a major theoretical review and concluded by a new synthesis informed by the research findings. Many pieces of empirical work contribute only marginally to the articulation of theory; a person interested in the role of elections in liberal democracies may undertake a great deal of empirical work into various aspects of how people make their voting decisions (for human geography, this implies a particular focus on locational aspects of those decisions: Johnston, 1985a), only occasionally using this material to develop theoretical understanding of liberal democracy (Johnston, 1984e, 1986e). Existence of a well-articulated and clearly specified theoretical perspective helps to ensure that the how questions are sensible ones, and the resulting descriptive analyses of value in later research.

Why, then, do we need empirical research? To some critics, especially those of a positivist orientation, empirical work seems to be irrelevant to a realist approach because: (a) realist theory can account, *a posteriori*, for all empirical outcomes, and so cannot be falsified; and (b) realist theory cannot predict, *a priori*, specific empirical outcomes, and so cannot be verified. It is supposed to help us to understand empirical and actual worlds, but lacks any criteria for evaluating its validity, or so it seems. Thus empirical work would appear to be superfluous (even counter-productive); like the normative theories of spatial science, it is right, irrespective of what is observed (see p. 54). Such an argument reflects a failure to appreciate the full realist case, and hence the role of empirical work. The case can be summarized as follows:

1   To organize physical reproduction, people have created modes of production, each of which has a number of economic imperatives within it (a base), and on which its survival depends;
2   The operation of those modes of production requires a set of social relations plus an institutional apparatus (a superstructure) within which those relations are organized;
3   Individuals learn how to live within the modes of production by their socialization into the institutional apparatus;
4   Daily life involves a continuous stream of decisions, taken freely by the individuals concerned but acting within the context of their institutional socialization; and
5   The consequences of those decisions become part of the context within which future action takes place.

We have, then, an economic base, a superstructure within which that economic base is activated, and individuals doing the activation. Base and superstructure are both human creations; superstructure is continually being modified by human action in order to maintain the structure as a whole, to ensure that the imperatives of the base are actualized and the whole mode of production does not collapse (which in capitalism means ensuring that accumulation continues). Theoretical research can illuminate the base for us, and can show why a superstructure is needed, but it cannot predict the contents of the superstructure in detail. Capitalism needs a state, for example, but it does not need a liberal democratic state (Johnston, 1986e); nor does it need a particular set of electoral cleavages (Taylor and Johnston, 1979; Taylor, 1985a) and a particular voting system (Bogdanor and Butler, 1983), and lead to particular voting habits (Crewe and Denver, 1985). Thus unless empirical research is conducted we cannot appreciate the robustness of the structure, the range of activity that it enables. Nor can we evaluate our theory properly. Realist theory may be neither verifiable nor falsifiable in a positivist sense, but to be of any value it must be coherent; the concepts that it provides must provide an understanding of, and explanation for, the observed events. The French Socialist Party introduced a proportional representation electoral system in 1984 as the best way of retaining power after the next election, as it interpreted the electoral situation, and therefore of steering the society in the way it wishes it to go: empirical work tells us this, it shows *how* the party acted, but it cannot aid our understanding of *why* France is a state, holds elections, has a socialist party etc.

Without explicit theory of the base and superstructure, empirical research may not address 'relevant' how questions. Without empirical research, theoretical development is arid, since it lacks direct links to the worlds of experience and events. Such links are crucial, as the next two sections of this chapter indicate, because without empirical research – historical as well as contemporary; of as many places as possible – we (a) fail to appreciate the full complexity of the societies and structures we have created and sustained and (b) fail to appreciate how change can be achieved. This is particularly important for human geographers, because many of the 'contingently related conditions' within which action takes place are spatially as well as historically localized. This variety can be used to support empiricist/ positivist and idealist views and a failure to uncover the underlying reality. Empirical research must be used not as an end in itself but as a way of illuminating the theory of society, of the real world of mechanisms that cannot be apprehended (as Cox and McCarthy, 1982, do). And compara- tive empirical research, by people with a full understanding of different places and times, will then reveal how people have created their worlds, are recreating their worlds, and can create new worlds: despite claims that 'there is no alternative', empirical research can, and must, show how destinies are created by people interpreting structures and making contexts, how contexts and destinies can be changed, and how structures can be changed. Empirical research is not voyeurism, it is sensitization.

ECONOMISM AND AUTONOMY

Many critics of the realist approach to human geography within social science espoused here focus on its stress on structures and on the primacy that it gives to economic factors as influences/determinants of human activity: see, for example, Duncan and Ley's (1982) charge that: 'intersection of human geography with structural marxism has led to a passive model of man that is conservative and results in an obfuscation of the processes by which human beings can and do change the world' (p. 54). My own presentation is certainly not passive; the hope, too, is that it is neither conservative nor obfuscatory.

The primacy of economic factors is sometimes in the eye of the reader rather than that of the writer, but a clear case can be made for this orientation. Survival and reproduction are the basic goals of all individuals, and societies develop to further those goals in particular ways. Thus a capitalist society is based on a particular relationship between labour and nature, set within its own relations of production. But capitalism extends those relations of production way beyond the reproduction of labour. Its dynamo is the drive to accumulate, and in order to expand accumulation most areas of life are infiltrated by the commodification process (as suggested in chapter 2), and further commodities are invented in order to create markets for anything that can be produced. As a result, the relationships between people become part of what Marx called commodity fetishism, and understanding social relationships requires analysts to probe beneath their empirical appearances to the economic forces that underpin them.

The influence of economic factors on social relations goes well beyond the direct commodification of interactions. Not only does the economic base penetrate the relationship between, say, a customer and a shopkeeper, it also infiltrates and conditions the relationships between individuals within families and households. Those relationships need not be expressed in monetary form. But they are part of the preparation for living in a capitalist society and of sustaining that society. Similarly, the institutions of the societal superstructure play 'preparing and sustaining roles', some perhaps more directly (educational systems, organised religions) than others.

In many parts of life, however, it is not that the social relations 'prepare and sustain' that is crucial – it is that they do not resist. The capitalist mode of production is extremely robust, and can tolerate a great range of activities within its shell. It does not *require* particular forms of social behaviour. Thus, for example, in many, if not all, societies there are deep cleavages that are not directly linked to the relations of production, those between bourgeoisie and proletariat. Many of these cleavages are ethnically based, linked to skin colour. Others are racially based, linked to socially created differences, such as language and religion. Those cleavages are not necessities of the capitalist mode of production, which does not mean that they are of no value to it. The ethnically and racially based policies of discrimination practised by South African whites under the ideology of *apartheid* are not necessary to the promotion of capitalism – though they clearly have helped – and when the policies have seemed to be harming accumulation, they have been altered; the superstructure has been changed. Similarly, the policies of discrimination in Northern Ireland founded on religious differences were not necessary to capitalist development there. What we see in both cases is the linking of particular attributes to the relations of production, by the people who put those relations in place, in order to advance their goals. Such interpretations did not harm the capitalist society, and so were 'permitted', not in any sense that the economic system allowed them to, but rather that the people who benefited from them did not find it necessary to dismantle them in order to promote their economic goals.

The two examples just given relate to salient features of the societies concerned, and illustrate the robustness and flexibility of capitalism with regard to the organization of the relations of production. Similar examples could be given for other aspects of the superstructure – of the state, for example. As argued in chapter 3, the state is a necessity to the capitalist mode of production. But it, too, has many degrees of freedom. It is required to promote and legitimate capitalism, and to maintain social cohesion, but it is not restricted to those activities alone. All actions of individuals in the state apparatus reflect their interpretations of how they should proceed in particular contexts (the Sheffield steel closure, again), and those interpretations are consequences of their personal and group socialization. Faced with declining industries, for example, parts of the state apparatus can promote policies designed to achieve one or more of: (1) restructuring the labour market to hasten modernisation, by providing training in new skills; (2) mopping up the negative consequences of restructuring, by welfare state support of the unemployed, for example; and (3) redistributing labour market opportunities so that no particular groups are disadvantaged (Davies, Mason and Davies, 1984). But not all of these actions need to be related (at least directly) to the promotion-legitimation-cohesion trilogy. Decisions to redistribute employment opportunities (to the young, to women etc.) may reflect individual preferences of state apparatus decision-makers that are separate from (if not independent of) the imperatives that they are activating. And other policy decisions – support for the arts and for a range of minority groups, for example – are entirely feasible within the robust flexibility of the mode of production, so long as they do not harm the state's major roles and are not the basis of electoral defeat.

All of these examples raise the important issue of *autonomy*; what freedom of action is available to individuals in the areas of social relations? The general answer is clearly a great deal. No precise answer can ever be given, of course, since the boundaries shift continually as capitalism itself changes all the time. For researchers, this means that understanding the limits to autonomy is an empirical, not a theoretical, task, yet the reasons for the empirical limits are only incompletely understood if they are not set in theoretical context. (The issue of autonomy for the state, and especially for the local state apparatus, is a topic of considerable interest at the present time – e.g. Clark and Dear, 1984; Boddy and Fudge, 1984 – not only as a research theme but also, as the volume edited by Boddy and Fudge illustrates, as a guide to political practice.)

For human geographers, the issue of autonomy is of particular interest because the separate interpretations of social relations reflected in apparently autonomous social practices are the result of socialization *in places*. The development of social practices proceeds slowly, as a consequence of interpersonal interactions, which are almost always, if not invariably, culture- and place-bound. The world is the complex mosaic that it is because of this. People have created societies, at a variety of scales, comprising both institutional arrangements necessary for the sustenance of the mode of production and those which provide for a richness of social life outside any direct link to the relations of production. Such arrangements are necessary, but their form is not prescribed. What people have created reflects what they thought was needed – which in turn reflects earlier interpretations. The 'collective wisdom' of their societies provides the context for their interpretations; variations in collective wisdom, in the provision of concepts with which to apply intellectual powers, provide separate, place-bound, sets of resources for creating the future. As the future unfolds, so the resources change, as people learn and record. They perceive other societies too, and learn from their experiences. All the time they are required to survive, to sustain capitalism (as I suggested in chapter 2, opting out is not an option). Thus their activities at the scale of experience (itself not fixed) are inextricably linked to the scale of reality (p. 67); a fundamental empirical task for human geography is to understand how worlds of experience are produced and reproduced within a global world-economy, mediated by the state and other ideologies. Central in all of this is the action of the individual. Places are made and remade by people, using their intellectual tools in the context of socially given (i.e. created by other individuals) concepts. Empirical research shows, by comparative case studies, how this occurs – it illustrates how people behave, how they create and recreate cultures in the context of an economic base.

KNOWLEDGE AND PRACTICE

The pursuit of knowledge is, for some of those involved (if not all), a self-sustaining activity, undertaken for the 'pure' benefits that it brings. But to many others, including many participants, knowledge is being sought not just for itself but because it is of use. As outlined in chapter 6, utility is most commonly defined in pragmatic terms, so that applied research is that which enables the immediate solution of technical problems. All knowledge is applicable, however, and all 'pure' science has an applied branch: empiricist/positivist science with technical/social control; humanistic science with the promotion of self- and mutual awareness; and realist science with emancipation. Thus, although the scientists may maintain the purity of their approach, and abdicate any responsibility for the uses to which their work is put, they cannot claim that science is separate from society; even if society does not direct the scientists in what to do, it can direct how what they have done can be used, for knowledge once created can never be obliterated.

Realization of this argument became widespread, and its consequences acute, with the weapons research, culminating with the development of thermonuclear weapons, during the Second World War. This creation of the means of social control – always the purpose of military research – was also the creation of the means of social destruction, if not eventually of social annihilation. For many natural scientists now, the potential uses of their work are so horrendous that the credibility of 'pure' research is much in question. The pursuit of science and the pursuit of political goals are closely harnessed.

For social scientists, similar conclusions can be drawn, though without the link to mass destruction (yet the genocide practised by Nazi politicians in Germany in the 1930s and 1940s was based on a misrepresentation of social scientific research and scholarship, itself not necessarily of a very high quality). Thus political utility of research findings, if not political practice, is an issue for social scientists too. As outlined in chapter 6, social science can be applied in three ways.

Each of these applications can be linked to a particular form of political practice. Applied human geography as technical/social control involves the use of research findings to structure the future by reproducing the present – using descriptions of the present to guide actions in the future. This necessarily involves reproducing the power relations of the present. It is essentially a perpetuation – perhaps spatially altered – of the status quo. Applied human geography as the promotion of awareness involves the use of research to heighten people's appreciation. It seeks to improve both the quality of life and the ability of people to accommodate complexity and variety, but it does nothing to tackle the sources of that complexity and variety. Whereas the first type of applied geography is conservative, this second type is liberal (perhaps romantically so) and it, too, is attached to no programme aimed at altering the status quo.

All human geography is necessarily applicable human geography if not applied human geography, therefore, and all human geography in the empiricist/positivist and humanistic moulds is applicable to the maintainance of the status quo (the first is what Harvey, 1973, calls a status quo application, whereas the latter is counter-revolutionary because despite its claims it lacks the potential to achieve substantial change). In both cases, the applications may be of material benefit, but will not alter the relations of production and the power disparities involved; people may be better off materially but have no more, and conceivably less, control over their own lives.

Applied human geography in the realist mould has as its goal emancipation, the uncovering of the base and superstructure of society so that individuals will appreciate fully the enormity of the exploitative and alienating relationships on which society is based. The political programme is therefore revolutionary, since the aim – which is that promoted by Harvey (1984) under the label of a 'people's geography' – is to convince people that they should remove the exploitative, alienating system and replace it by one based on relationships among equals. Thus the purpose of research, or the use to which it is put, is to enhance realistic understanding, to enable people to link their worlds of empirical experiences and actual events to that of real mechanisms, hence the need for the symbiosis of empirical and theoretical research.

The political programme of revolution linked to realist research is necessarily a gradualist one, since the goal is to encourage people to realize their situation and to replace it; to convince workers of the world to unite, since they have nothing to lose but their chains. There are problems here, however. First, the empirical research conducted within a realist framework can nevertheless be applied positivistically – good description of the present can be used to reproduce the present, even if the describer intends otherwise. Secondly, the emancipation programme will have to counter the arguments for the status quo and will undoubtedly lack the centre-stage position occupied by the ideological forces being attacked. Thirdly, those ideological forces will marshal arguments based on the assumed superiority of the status quo, and the use of research findings in that context will itself have to be unmasked. Fourthly, and perhaps most important of all, mobilization will be extremely difficult with a majority of the world population. The classic marxist argument is: (1) that capitalism will solve all the technical problems of maintaining a high material standard of living, but only through alienating labour in order to support accumulation; (2) in response to this alienation, socialism will slowly be used to replace the need for accumulation by transferring the ownership of the means of production, distribution and exchange to the state and thereby removing the alienating relations of production and mode of organization in which the development:underdevelopment dualism is fundamental; and (3) the state will then slowly wither away, to be replaced by communism and an entirely collective mode of production based on anarchism, with high material standards of living but an absence of commodification and production-consumption for profit only, and without social relations built on commodity fetishism. Thus capitalism is a necessary precursor to socialism and communism, its overthrow coming about because it contains the seeds of its own destruction, seeds that must be carefully nurtured by the programme of emancipation. Unfortunately, the fruits of capitalism have yet to be experienced by most of the world's population, and many of them perceive the attraction of such fruits as much greater than the totalitarianism imposed in those states that have sought to build communism from feudalism or the early stage of capitalism only, and without the material foundation.

The programme of emancipation is a difficult one to implement, therefore, and the other two applied geographies can appear much more attractive : promotion of emancipation via the type of research commended here will, of itself, improve mutual awareness. Yet, if capitalism necessarily (a) is built on a development: underdevelopment dualism that is spatially structured (Browett, 1984, suggests that spatial inequality is not a necessity of the capitalist mode of production; N. Smith, 1986, argues that it is); and (b) passes through a sequence of boom and crises, with the latter involving both spatial restructuring and the potential for global conflict (Harvey, 1985; O'Loughlin, 1986), then a successful articulation of theoretical understanding and empirical research will demonstrate that, for a majority of people, the benefits of the status quo are necessarily transitory.

All human geography is applicable, therefore; all human geography is linked to political practice. The programme of political practice to which my human geography is applied, by me, is emancipation – though I realize fully that much of what I do empirically is applicable by others (and perhaps indirectly by me) as either or both of technical control or awareness-increasing. (Neither is necessarily 'bad'. Technical control applications may benefit people materially – and perhaps thereby hasten the revolution! Mutual awareness will almost certainly be beneficial in the short term, but needs to be harnessed to realistic understanding in the long term.) Pursuit of my programme involves the linked development of theory-articulation with empirical investigation and the evolution of educational strategies that will promote emancipation. As made clear in chapter 8, this is extremely difficult (in many pessimistic moments I conclude either or both of its impossibility and my unsuitedness for it). But to me it is the logical outcome of the understanding that my realist approach imparts.

One of the most frequently cited sentences from Marx's voluminous writings is his claim that:

Men make their own history, but they do not make it under circumstances chosen by themselves but under circumstances directly encountered, given and transmitted from the past.

This has been paraphrased in a great variety of ways, including:

People make their own geography, but not in circumstances of their own choosing.

In this the word geography can be used in both the vernacular and the academic sense. With regard to the vernacular, it states that we are socialized into particular spatial structures and interpretations of nature, within which we are enabled to produce our own, according to our decisions, but constrained by that context and the interpretive powers that are given to us. With regard to the academic, it states that we are socialized into a particular division of labour which enables us to work in particular ways, according to our choice, but also constrains our thinking by the concepts that we are given.

I accept both interpretations. They provide me with a disciplinary focus on an important component of the constitution of society. We make vernacular geographies in the localized contexts of those that others have made, and we undertake academic geographies to explicate how that manufacture proceeds. But those localized contexts are based on interpretations of a general, structural, context that we have also produced – the capitalist mode of production. Our vernacular geographies incorporate such interpretations. Our academic geographies seek to explicate why that structural context is present, and why it directs our actions in particular ways. We seek, as academic geographers, to understand how and why we act, thereby providing us with both an understanding of the constitution of society and the knowledge with which we can fashion a new society. Our goals are understanding and application, with the latter to be achieved via education and emancipation, by our providing the conceptual resources with which people will appreciate both how we have produced our present situation and how we can act positively in the creation of the future.

The existence of human geography within an academic division of labour may not be the optimal path to emancipation and the creation of a future. But it exists and, for Britain at least, Harvey (1984) is wrong in his claim that 'academic geography failed to build a position of power, prestige, and respectability within the academic division of labour' (p. 4). It is a ready-made vehicle for the promotion of an emancipatory programme of research and teaching, and it is towards that end that our efforts should be directed.

Bibliography

Abler, R. F., Adams, J. S. and Gould, P. R. 1970: *Spatial Organization: The Geographer's View of the World*. Englewood Cliffs, Prentice-Hall.

Amedeo, D. and Golledge, R. G. 1975: *An Introduction to Scientific Reasoning in Geography*. New York: John Wiley.

Archer, J. C. and Taylor, P. J. 1981: *Section and Party: A Political Geography of American Presidential Elections from Andrew Jackson to Ronald Reagan*. Chichester: John Wiley.

Ashby, J. H. 1981: The influence of publishers on academia. *Environment and Planning A*, 13, 1175–6.

Baker, A. R. H. and Gregory, D. 1984: Some terrae incognitae in historical geography: an exploratory discussion. In Baker, A. R. H. and Gregory, D. (eds) *Explorations in Historical Geography*. Cambridge: Cambridge University Press, 180–94.

Balchin, W. G. V. 1981: Book review. *Geography*, 66, 255–6.

Batty, M. 1976: *Urban Modelling: Algorithms, Calibrations, Predictions*. Cambridge: Cambridge University Press.

Bennett, R. J. 1974: Process identification for time series modelling in urban and regional planning. *Regional Studies*, 8, 157–74.

—— 1981: Quantitative and theoretical geography in western Europe. In Bennett, R. J. (ed.) *European Progress in Spatial Analysis*. London: Pion, 1–32.

—— 1985: A reappraisal of the role of spatial science and statistical inference in geography in Britain. *L'Espace Geographique*, 14, 23–8.

—— and Chorley, R. J. 1978: *Environmental Systems: Philosophy, Analysis and Control*. London: Methuen.

Bernstein, R. J. 1985: *Habermas and Modernity*. Cambridge: Cambridge University Press.

Berry, B. J. L. 1970: City size and economic development. In Jakobson, L. and Prakesh, V. (eds) *Urbanization and National Development*. Beverly Hills: Sage Publications, 111–56.

—— 1973: *The Human Consequences of Urbanization*. London: Macmillan.

—— and Marble, D. F. 1968: Introduction. In Berry, B. J. L. and Marble, D. F. (eds) *Spatial Analysis: A Reader in Statistical Geography*. Englewood Cliffs: Prentice-Hall, 1–9.

Betjeman, Sir John (1978): Postcript for Highworth. In Guest, J., *The Best of Betjeman*. London: Penguin.

Billinge, M., Gregory, D. and Martin, R. L. (eds) 1984: *Reflections on a Revolution*. London: Macmillan.

Blainey, G. 1966: *The Tyranny of Distance*. Melbourne: Sun Books.

Boddy, M. and Fudge, C. (eds) 1984: *Local Socialism: Labour Councils and New Left Alternatives*. London: Macmillan.

Bogdanor, V. and Butler, D. (eds) 1983: *Democracy and Elections: Electoral Systems and their Political Consequences*. Cambridge: Cambridge University Press.

Bordessa, R. and Bunge, W. 1975: *The Canadian Alternative*. Downsview, Ont.: Geographical Monographs, York University.

Bosque-Sendra, J., Rodrigues, V. and Santos, J. H. 1983: Quantitative geography in Spain. *Progress in Human Geography*, 7, 370–85.

Braithwaite, R. B. 1960: *Scientific Explanation*. New York: Harper and Row.

Browett, J. 1984: On the necessity and inevitability of uneven spatial development under capitalism. *International Journal of Urban and Regional Research*, 8, 155–76.

Brunsden, D. S. 1985: Voice for Geography. *The Geographical Magazine*, 57, 464–5.

Bunge, W. 1971: *Fitzgerald: Geography of a Revolution*. Cambridge, MA: Schlenkman.

Burton, I., Kates, R. W. and White, G. F. 1978: *The Environment as Hazard*. Oxford: Oxford University Press.

Butler, R. A. 1980: Introduction to the Penguin edition of B. Disraeli, *Sybil*. London: Penguin.

Buttimer, A. 1974: *Values in Geography*. Washington DC: Association of American Geographers.

—— 1984: *The Practice of Geography*. London: Longman.

Carter, H. 1982: Review of *City and Society*. *Progress in Human Geography*, 6, 303–6.

Castells, M. 1983: *The City and the Grassroots*. London: Edward Arnold.

Clark, C. 1985: Geography must be alive and kicking. *Area*, 17, 174–5.

Clark, G. L. and Dear, M. J. 1984: *State Apparatus: Structures of Language and Legitimacy*. London: George Allen and Unwin.

Clayton, K. M. 1985: The state of geography. *Transactions, Institute of British Geographers*, NS10, 5–16.

—— and O'Riordan, T. 1977: The readership of *Transactions* and the role of the IBG. *Area* 9, 96–100.

Coffey, W. J. 1981: *Geography: Towards a General Spatial Systems Approach*. London: Methuen.

Coleman, A. 1985: *Utopia on Trial*. London: H. Wiseman.

Collingwood, R. G. 1965: *Essays in the Philosophy of History*. Austin: University of Texas Press.

Cooke, R. U. 1985: Applied geomorphology. In A. Kent (ed.) *Perspectives on a Changing Geography*. Sheffield: The Geographical Association, 36–47.

Cox, N. J. and Jones, K. 1981: Exploratory data analysis. In Wrigley, N. and Bennett, R. J. (eds) *Quantitative Geography: a British View*. London: Routledge and Kegan Paul, pp. 135–43.

Cox, K. R. 1976: American geography: social science emergent. *Social Science Quarterly*, 57, 182–207.

—— and McCarthy, J. J. 1982: Neighbourhood activism as a politics of turf: a critical analysis. In Cox, K. R. and Johnston, R. J. (eds) *Conflict, Politics and the Urban Scene*. London: Longman, 196–219.

Crewe, I. and Denver, D. (eds) 1985: *Electoral Change in Western Democracies: Patterns and Sources of Electoral Volatility*. London: Croom Helm.

Davie, M. R. 1938: The pattern of urban growth. In Murdoch, G. P. (ed.) *Studies in the Science of Society*. New Haven: Yale University Press, 131–61.

Davies, T., Mason, C. and Davies, L. 1984: *Government and Local Labour Market Policy Implementation*. Aldershot: Gower Press.

Downs, A. 1957: *An Economic Theory of Democracy*. New York: Harper and Row.

Duncan, J. S. and Ley, D. 1982: Structural marxism and human geography: a critical assessment. *Annals of the Association of American Geographers*, 72, 30–59.

Dunleavy, P. 1979: The urban basis of political alignment. *British Journal of Political Science*, 9, 409–43.

—— 1981: Perspectives on urban studies. In Blowers, A. T., Brook, C., Dunleavy, P. and McDowell, L. (eds) *Urban Change and Conflict: an Interdisciplinary Reader*. London: Harper and Row, 1–16.

—— and Husbands, C. T. 1985: *British Democracy at the Crossroads: Voting and Party Competition in the 1980s*. London: George Allen and Unwin.

Elazar, D. J. 1966: *American Federalism: The View from the States*. New York: T. Y. Crowell.

Eliot Hurst, M. E. 1980: Geography, social science and society: towards a de-definition. *Australian Geographical Studies*, 18, 3–21.

—— 1985: Geography has neither existence nor future. In Johnston, R. J. (ed) *The Future of Geography*. London: Methuen, 55–91.

Elliott, W. E. Y. 1974: *The Rise of Guardian Democracy*. Cambridge, MA: Harvard University Press.

Eyles, J. 1985: *Sense of Place*. Warrington: Silverwood Press.

Finer, S. E. (ed.) 1975: *Adversary Politics and Electoral Reform*. London: Anthony Wigram.

Fothergill, S. and Vincent, J. 1985: *The State of the Nation Atlas*. London: Pan Books.

Gale, S. 1982: Notes on an institutionalist approach to geography: two-dimensional man in a two-dimensional society. In Flowerdew, R. (ed.) *Institutions and Geographical Patterns*. London: Croom Helm, 51–73.

Gamble, A. M. and Walkland, S. A. 1984: *The British Party System and Economic Policy, 1945–1983*. Oxford: Oxford University Press.

Giddens, A. (ed.) 1974: *Positivism and Sociology*. London: Heinemann.

_____ 1983: *A Critique of Contemporary Historical Materialism*. London: Macmillan.

_____ 1984: *The Construction of Society*. Oxford: Polity Press.

Gill, D. 1982: *Geography for the Young School Leaver: A Critique*. Working Paper 2, Centre for Multicultural Education, University of London Institute of Education.

Golledge, R. G. 1980: A behavioral view of mobility and migration research. *The Professional Geographer*, 33, 247–51.

_____ 1982: Fundamental conflicts and the search for geographical knowledge. In P. R. Gould and G. Olsson (eds) *A Search for Common Ground*. London: Pion, pp. 11–23.

Goodson, I. F. 1981: Becoming an academic subject: patterns of explanation and evolution. *British Journal of Sociology of Education*, 2, 163–79.

_____ 1985: Subjects for study. In I. F. Goodson (ed.) *Social Histories of the Secondary Curriculum*. Lewes: The Falmer Press, 343–68.

Goudie, A. S. 1982: Should human and physical geography go their separate ways? *Graticule* (Journal of Queen's University Belfast Geographical Society), 25, 25–6.

Gould, P. R. 1979: Geography 1957–1977: the Augean period. *Annals of the Association of American Geographers*, 69, 139–51.

_____ 1985: *The Geographer at Work*. London: Routledge and Kegan Paul.

Graf, W. L., Trimble, S. W., Toy, T. J. and Costa, J. E. 1980: Geographic geomorphology in the eighties. *The Professional Geographer*, 32, 279–84.

Gregory, D. 1978: *Ideology, Science and Human Geography*. London: Hutchinson.

_____ 1980: The ideology of control: systems theory and geography. *Tijdschrift voor Economische en Sociale Geografie*, 71, 327–42.

_____ 1981: Human agency and human geography. *Transactions, Institute of British Geographers*, NS6, 1–18.

_____ and Urry, J. (eds) 1985: *Social Relations and Spatial Structures*. Macmillan: London.

Gregory, K. J. 1985: *The Nature of Physical Geography*. London: Edward Arnold.

_____ and Williams, R. F. 1981: Physical geography from the newspaper. *Geography*, 66, 42–52.

Gregory, S. 1978: The role of physical geography in the curriculum. *Geography*, 63, 251–64.

Gregson, N. 1986: On duality and dualism: the case of structuration and time geography. *Progress in Human Geography*, 10.

Guelke, L. 1974: An idealist alternative in human geography. *Annals of the Association of American Geographers*, 64, 193–202.

_____ 1977: Geography and logical positivism. In Herbert, D. T. and Johnston, R. J. (eds) *Geography and the Urban Environment*, vol. 1. Chichester: John Wiley, 35–61.

____ 1982: *Historical Understanding in Geography: An Idealist Approach.* Cambridge: Cambridge University Press.

Guest, J. 1978: *The Best of Betjeman.* London: Penguin.

Haggett, P. 1965: *Locational Analysis in Human Geography.* London: Edward Arnold.

____ and Chorley, R. J. 1969: *Network Models in Geography.* London: Edward Arnold.

____ , Cliff, A. D. and Frey, A. E. 1977: *Locational Analysis in Human Geography* (2nd edn). London: Edward Arnold.

Hall, P. 1980: *Great Planning Disasters.* London: Penguin.

Harries, K. D. 1976: Observations on radical versus liberal theories of crime causation. *The Professional Geographer*, 28, 110–13.

Harrison, P. 1983: *Inside the Inner City: Life Under the Cutting Edge.* London: Penguin.

Harrison, R. T. and Livingstone, D. N. 1980: Philosophy and problems in human geography: a presuppositional approach. *Area*, 12, 25–32.

____ and Livingstone, D. N. 1982: Understanding in geography: structuring the subjective. In Herbert, D. T. and Johnston, R. J. (eds) *Geography and the Urban Environment* vol. 5. Chichester: John Wiley, 1–140.

Hart, J. F. 1982: The highest form of the geographer's art. *Annals of the Association of American Geographers*, 72, 1–29.

Hartshorne, R. 1959: *Perspective on the Nature of Geography.* Chicago: Rand McNally.

____ 1984: in *The Geographical Journal*, 150, 429.

Harvey, D. 1969: *Explanation in Geography.* London: Edward Arnold.

____ 1973: *Social Justice and the City.* London: Edward Arnold.

____ 1974a: What kind of geography for what kind of public policy? *Transactions, Institute of British Geographers*, 63, 18–24.

____ 1974b: Class-monopoly rent, finance capital, and the urban revolution. *Regional Studies*, 8, 239–55.

____ 1975: The political economy of urbanization in advanced capitalist societies. In Gappert, G. and Rose, H. M. (eds) *The Social Economy of Cities.* Beverly Hills: Sage Publications, 119–63.

____ 1982: *The Limits to Capital.* Oxford: Basil Blackwell.

____ 1984: On the history and present condition of geography: an historical materialist manifesto. *The Professional Geographer*, 36, 1–11.

____1985: The geopolitics of capitalism. In Gregory, D. and Urry, J. (eds) *Space and Social Structures.* London: Macmillan, 128–63.

Hay, A. M. 1979: Positivism in human geography: response to critics. In Herbert, D. T. and Johnston, R. J. (eds) *Geography and the Urban Environment*, vol. 2. Chichester: John Wiley, 1–26.

____ and Johnston, R. J. 1979: Search and the choice of shopping centre: two models of variability in destination selection. *Environment and Planning A*, 11, 791–804.

_____ and Johnston, R. J. 1980: Spatial variations in grocery prices: further attempts at modelling. *Urban Geography*, 1, 189–200.

Herbert, D. T. 1979: Introduction: geographical perspectives and urban problems. In Herbert, D. T. and Smith, D. M. (eds) *Social Problems and the City*. Oxford: Oxford University Press, 1–9.

Huckle, J. 1983: The politics of school geography. In Huckle, J. (ed.) *Geographical Education: Reflection and Action*. Oxford: Oxford University Press, 143–54.

Hudson, R. 1983: The question of theory in political geography: outlines for a critical theory approach. In Kliot, N. and Waterman, S. (eds) *Pluralism and Political Geography: People, Territory and State*. London: Croom Helm, 29–35.

Husbands, C. T. 1983: *Racial Exclusionism in the City*. London: George Allen and Unwin.

Jackson, P. 1985: Urban ethnography. *Progress in Human Geography*, 9, 157–76.

_____ and Smith, S. J. 1981: Introduction. In Jackson, P. and Smith, S. J. (eds) *Social Interaction and Ethnic Segregation*. London: Academic Press, 1–18.

_____ and Smith, S. J. 1984: *Exploring Social Geography*. London: George Allen and Unwin.

Johnston, R. J. 1976: Observations on accounting procedures and urban-size policies. *Environment and Planning A*, 8, 327–40.

_____ 1980a: On the nature of explanation in human geography. *Transactions, Institute of British Geographers*, NS5, 402–12.

_____ 1980b: *City and Society: An Outline for Urban Geography*. London: Penguin.

_____ 1981: Applied geography, quantitative analysis, and ideology. *Applied Geography*, 1, 213–19.

_____ 1982: *The American Urban System: A Geographical Perspective*. New York: St Martin's Press.

_____ 1983a: *Geography and Geographers: Anglo-American Human Geography since 1945*. London: Edward Arnold.

_____ 1983b: *Philosophy and Human Geography: An Introduction to Contemporary Approaches*. London: Edward Arnold.

_____ 1983a: Resource analysis, resource management and the integration of physical and human geography. *Progress in Physical Geography*, 7, 127–46.

_____ 1984a: The region in twentieth century British geography. *History of Geography Newsletter*, 4, 26–35.

_____ 1984b: The feedback component of the pork-barrel: tests using the results of the 1983 general election in Britain. *Environment and Planning A*, 15, 1567–1716.

_____ 1984c: The world is our oyster. *Transactions, Institute of British Geographers*, NS9, 443–59.

_____ 1984d: *Residential Segregation, the State and Constitutional Conflict in American Urban Areas*. London: Academic Press.

_____ 1984e: The political geography of electoral geography. In Taylor, P. J. and House, J. W., (eds) *Political Geography: Recent Advances and Future Directions*. London: Croom Helm, 133–48.

_____ 1985a: *The Geography of English Politics: The General Election of 1983*. London: Croom Helm.

_____ 1985b: To the ends of the earth. In Johnston, R. J. (ed.) *The Future of Geography*. London and New York: Methuen, 326–38.

_____ 1985c: The neighbourhood effect revisited: spatial science or political socialisation. *Environment and Planning D: Society and Space*, 3.

_____ 1986a: Places matter. *Irish Geography*, 18, 58–63.

_____1986b: Philosophy, ideology, and geography. In Gregory, D. and Walford, R. (eds) *New Directions in Geography*. London: Macmillan.

_____ 1986c: Understanding and solving American urban problems: geographical contributions? *The Professional Geographer*, 38.

_____ 1986d: Perspectives on applied human geography. In J. Bosque Maurel (ed.) *Homenage a Manuel de Teran Alvarez*. Madrid: Universidad Complutense.

_____ 1986e: Individual freedom and the world-economy. In Johnston, R. J. and Taylor, P. J. (eds) *A World in Crisis? Geographical Perspectives*. Oxford: Basil Blackwell.

_____ and Claval, P. (eds) 1984: *Geography since the Second World War: An International Survey*. London: Croom Helm.

_____ and Gregory, S. 1984: The United Kingdom. In Johnston, R. J. and Claval, P. (eds) *Geography since the Second World War: An International Survey*. London: Croom Helm, 107–31.

_____ and Hay, A. M. 1979: Variability in grocery prices. *Area*, 11, 160–2.

_____ , O'Neill, A. B. and Taylor, P. J. 1985: *The Geography of Party Support: Comparative Studies in Electoral Stability*. Seminar Paper 40, Department of Geography, University of Newcastle upon Tyne.

Joseph, Rt Hon. Sir Keith 1985: *Geography in the School Curriculum*. Sheffield: The Geographical Association.

Keat, R. 1981: *The Politics of Social Theory*. Oxford: Basil Blackwell.

Kidron, M. and Segal, R. 1984: *The New State of the World Atlas*. London: Pan Books.

King, L. J. and Golledge, R. G. 1978: *Cities, Space and Behavior*. Englewood Cliffs: Prentice-Hall.

Kirk, W. 1963: Problems of geography. *Geography*, 48, 357–71.

Kolakowski, L. 1972: *Positivist Philosophy*. London: Penguin.

Knorr-Cetina, K. 1982: Scientific communities or trans-epistemic arenas of research – a critique of quasi-economic models of science. *Social Studies of Science*, 12, 101–30.

van der Laan, L. and Piersma, A. 1982: The image of man: paradigmatic cornerstone in human geography. *Annals of the Association of American Geographers*, 74, 411–26.

Lacey, A. R. 1976: *A Dictionary of Philosophy*. London: Routledge and Kegan Paul.

Lakatos, I. 1978: Falsification and the methodology of scientific research programmes. In Worrall, J. and Currie, G. (eds) *The Methodology of Scientific*

Research Programmes, Philosophical Papers I. Cambridge: Cambridge University Press, 1–101.

Langton, J. 1984: Potentialities and problems of adopting a systems approach to the study of change in human geography. In Board, C., Chorley, R. J., Haggett, P. and Stoddart, D. R. (eds) *Progress in Geography, 4.* London: Edward Arnold, 125–80.

Ley, D. 19 74: *The Black Inner City as Frontier Outpost: Images and Behavior of a Philadelphian Neighborhood.* Washington DC: Association of American Geographers.

____ 1977: Social geography and the taken-for-granted world. *Transactions, Institute of British Geographers*, NS2, 498–512.

____ 1980: Liberal ideology and the post-industrial city. *Annals of the Association of American Geographers*, 70, 238–58.

____ 1983: *A Social Geography of the City.* New York: Harper and Row.

Lichtenberger, E. 1984: The German-speaking countries. In Johnston, R. J. and Claval, P. (eds) *Geography since the Second World War: An International Survey.* London: Croom Helm, 156–84.

Lipset, S. M. and Rokkan, S. E. 1967: Cleavage structures, party systems and voter alignments: an introduction. In Lipset, S. M. and Rokkan, S. E. (eds) *Party Systems and Voter Alignments.* New York: The Free Press, 3–64.

Lukács, G. 1923: *History and Class Consciousness.* London: Molin Press (tr. R. Livingstone, 1971).

McDowell, L. and Massey, D. 1984: Women's place. In Massey, D. and Allen, J. (eds) *Geography Matters!* Cambridge: Cambridge University Press.

Mann, M. 1984: The autonomous power of the state: its origins, mechanisms, and results. *European Journal of Sociology*, 25, 185–213.

Marshall, J. U. 1985: Geography as a scientific enterprise. In Johnston, R. J. (ed.) *The Future of Geography.* London: Methuen, 113–28.

Massey, D. and Allen, J. (eds) 1984: *Geography Matters!* Cambridge: Cambridge University Press.

Mercer, D. C. 1978: Revolutionary geography. *Progress in Human Geography*, 2, 537–42.

____ 1985a: *Reading the Book of Nature: Physical and Human Geography and the Limits of Science.* Working Paper 18, Department of Geography, Monash University, Clayton, Victoria.

____ 1985b: On Marx, morals, method and meaning. *Australian Geographical Studies*, 23, 139–57.

Miller, W. L. 1977: *Electoral Dynamics.* London: Macmillan.

Morrill, R. L. 1970: *The Spatial Organization of Society.* Belmont, CA: Wadsworth.

NAS/NRC 1965: *The Science of Geography.* Washington DC: National Academy of Sciences/National Research Council.

Nystuen, J. D. 1968: Identification of some fundamental spatial concepts. In Berry, B. J. L. and Marble, D. F. (eds) *Spatial Analysis.* Englewood Cliffs: Prentice-Hall, 35–41.

Oakeshott, M. 1983: *On History*. Oxford: Basil Blackwell.

O'Loughlin, J. 1986: Hegemonic competition and local conflicts in the Third World. In Johnston, R. J. and Taylor, P. J. (eds) *A World in Crisis? Geographical Perspectives*. Oxford: Basil Blackwell.

Olsson, G. 1982: –/– In Gould, P. R. and Olsson, G. (eds) *A Search for Common Ground*. London: Pion, 223–31.

Openshaw, S. and Taylor, P. J. 1981: The modifiable areal unit problem. In Wrigley, N. and Bennett, R. J. (eds) *Quantitative Geography: A British View*. London: Routledge and Kegan Paul, 60–70.

O'Riordan, T. 1977: Environmental ideologies. *Environment and Planning A*, 9, 3–14.

Orme, A. R. 1980: The need for physical geography. *The Professional Geographer*, 32, 141–9.

Park, R. E. 1926: The urban community as a spatial pattern and a moral order. In Burgess, E. W. (ed.) *The Urban Community*. Chicago: University of Chicago Press.

Peet, J. R. 1982: International capital, international culture. In Taylor, M. J. and Thrift, N. J. (eds) *The Geography of Multinationals*. London: Croom Helm, 275–302.

—— 1985: An introduction to Marxist geography. *Journal of Geography*, 84, 5–10.

—— 1986: The destruction of regional cultures. In Johnston, R. J. and Taylor, P. J. (eds) *A World in Crisis? Geographical Perspectives*. Oxford: Basil Blackwell.

Pepper, D. 1983: Bringing physical and human geographers together: why is it so difficult? In London Group of Union of Socialist Geographers, *Society and Nature*. London, 19–31.

—— 1984: *The Roots of Modern Environmentalism*. London: Croom Helm.

Philbrick, A. K. 1957: Principles of areal functional organization in regional human geography. *Economic Geography*, 33, 299–336.

Phillips, D. and Williams, A. 1984: *Rural Britain: A Social Geography*. Oxford: Basil Blackwell.

Pickles, J. 1985: *Phenomenology, Science and Geography: Spatiality and the Human Sciences*. Cambridge: Cambridge University Press.

Pocock, D. C. D. 1981: Introduction: imaginative literature and the geographer. In Pocock, D. C. D. (ed.) *Humanistic Geography and Literature: Essays on the Experience of Place*. London: Croom Helm, 9–19.

Pred, A. R. 1967: *Behavior and Location I*. Lund: C. W. K. Gleerup.

—— 1984a: Place as historically contingent process: structuration and the time-geography of becoming places. *Annals of the Association of American Geographers*, 74, 279–97.

—— 1984b: Structuration, biography formation, and knowledge: observations on port growth during the late mercantile period. *Environment and Planning D: Society and Space*, 2, 251–76.

Ramphal, S. S. 1985: A world turned upside down. *Geography*, 70, 193–205.

Robertson, D. 1984: *Class and the British Electorate*. Oxford: Basil Blackwell.

Rose, C. 1981: Wilhelm Dilthey's philosophy of historical understanding: a neglected heritage of contemporary humanistic geography. In Stoddart, D. R. (ed.) *Geography, Ideology and Social Concern*. Oxford: Basil Blackwell, 99–133.

Rushton, G. 1969: Analysis of spatial behavior by revealed space preferences. *Annals of the Association of American Geographers*, 59, 391–400.

Sack, R. D. 1974: The spatial separatist theme in geography. *Economic Geography*, 50, 1–19.

―― 1980: *Conceptions of Space in Social Thought: A Geographic Perspective*. London: Macmillan.

―― 1983: Human territoriality: a theory. *Annals of the Association of American Geographers*, 73, 55–74.

―― 1986: *Human Territoriality*. Cambridge: Cambridge University Press.

Sayer, A. 1979: Epistemology and conceptions of people and nature in geography. *Geoforum*, 10, 19–44.

―― 1982: Explanation in economic geography. *Progress in Human Geography*, 6, 68–88.

―― 1983: Notes on geography and the relationship between people and nature. In The London Group of the Union of Socialist Geographers, *Society and Nature*. London, 47–57.

―― 1984: *Method in Social Science: A Realist Approach*. London: Hutchinson.

―― 1985: The difference that space makes. In D. Gregory and J. Urry (eds) *Social Relations and Spatial Structures*. London: Macmillan, 49–66.

Scarbrough, E. 1984: *Political Ideology and Voting: An Exploratory Study*. Oxford: The Clarendon Press.

Schaefer, F. K. 1953: Exceptionalism in geography: a methodological examination. *Annals of the Association of American Geographers*, 43, 226–49.

Schattschneider, E. E. 1960: *The Semi-Sovereign People*. New York: Harper and Row.

Sennett, R. 1970: *The Uses of Disorder*. London: Penguin.

Short, J. R. 1984: *The Urban Arena: Capital, State and Community in Contemporary Britain*. London: Macmillan.

Silk, J. 1982: Commentary on 'On the nature of explanation in human geography'. *Transactions, Institute of British Geographers*, NS7, 380–4.

Simmons, I. G. and Cox, N. J. 1985: Holistic and reductionist approaches to geography. In Johnston, R. J. (ed.) *The Future of Geography*. London: Methuen, 43–58.

Smith, D. M. 1977: *Human Geography: A Welfare Approach*. London: Edward Arnold.

Smith, N. 1985: On the necessity of uneven development. *International Journal of Urban and Regional Research*, 10.

de Souza, A. R. 1983: Talks with teachers: John Fraser Hart. *Journal of Geography*, 82, 54–8.

Stoddart, D. R. (ed.) 1981: *Geography, Ideology and Social Concern.* Oxford: Basil Blackwell.

Sutherland, J. 1972: Introduction to the Penguin Edition of A. Trollope, *Phineas Finn.* London: Penguin.

Taylor, P. J. and Johnston, R. J. 1979: *Geography of Elections.* London: Penguin.

—— 1981a: Geographical scales within the world economy approach. *Review*, 5(1), 3–11.

—— 1981b: Factor analysis in geographical research. In Bennett, R. J. (ed.) *European Progress in Spatial Analysis.* London: Pion, 251–67.

—— 1983: The question of theory in political geography. In Kliot, N. and Waterman, S. (eds) *Pluralism and Political Geography: People, Territory and State.* London: Croom Helm, 9–18.

—— 1985a: *Political Geography: World–Economy, Nation-State and Locality.* London: Longman.

—— 1985b: The value of a geographical perspective. In R. J. Johnston (ed.) *The Future of Geography.* London: Methuen, 92–110.

—— 1986a: Chaotic conceptions, antinomies, dilemmas and dialectics: who's afraid of the capitalist world economy. *Political Geography Quarterly*, 5.

—— 1986b: The myth of developmentalism. In Gregory, D. and Walford, R. (eds) *New Horizons in Geography.* London: Macmillan.

Thomas, K. 1984: *Man and the Natural World: Changing Attitudes in England 1500–1800.* London: Penguin.

Thrift, N. J. 1983: On the determination of social action in space and time. *Environment and Planning D: Society and Space*, 1, 23–57.

—— 1985: The state of British urban and regional research in a time of economic crisis. *Environment and Planning A*, 17, 7–24.

Trow, M. 1 976: The American academic department as a context for learning. *Studies in Higher Education*, 1, 11–22.

Urwin, D. W. and Rokkan, S. E. 1983: *Economy, Territory, Identity: Politics of West European Peripheries.* London: Sage Publications.

Wallerstei n, I. 1979: *The Capitalist World-Economy.* Cambridge: Cambridge University Press.

Williams, R. 1958: *Culture and Society 1780–1950.* London: Chatto and Windus.

—— 1973: *The Country and the City.* London: Chatto and Windus.

Williamson, J. G. 1965: Regional income inequality and the process of national development: a description of the patterns. *Economic Development and Cultural Change*, 13, 3–84.

Wilson, A. G. 1974: *Urban and Regional Models in Geography and Planning.* Chichester: John Wiley.

Wirth, E. 1984: Geografie als moderne theorieorientierte Sozialwissenschaft. *Erdkunde*, 38 78–8. [An English-language version is published as 'The abolition of man in modern social science orientated geography', Geography Department, King's College, London, Occasional Paper 23, 1984.]

Wise, M. J. 1984: World geography since the war. *The Geographical Magazine*, December, 651.

Woldenberg, M. J. and Berry, B. J. L. 1967: Rivers and central places: analogous systems? *Journal of Regional Science*, 7(2), 129–39.

Wooldridge, S. W. and East, W. G. 1958: *The Spirit and Purpose of Geography*. London: Hutchinson.

Worsley, P. 1979: Whither geomorphology? *Area*, 11, 97–101.

―― 1985: Physical geography and the natural environmental sciences. In R. J. Johnston (ed.) *The Future of Geography*. London: Methuen, 27–42.

Wrigley, N. 1985: *Categorical Data Analysis for Geographers and Environmental Scientists*. London and New York: Longman.

Yeates, M. H. 1983: Review of *The American Urban System. The Professional Geographer*, 35, 504–5.

Zelinsky, W. 1975: The demigod's dilemma. *Annals of the Association of American Geographers*, 65, 173–43.

Index

Abler, R. F. 84, 182
abstraction 63, 64, 124, 142, 170
accumulation 26–7, 30–1, 34, 35, 44, 46, 101, 134
adversary politics 16–17
alienation 11
Amedeo, D. 85, 182
anarchism 2, 61, 164
applied geography 4–5, 100–21, 159, 177–81
Archer, J. C. 98, 182
Ashby, J. H. 166
Association of American Geographers 166
autobiography 2, 3
autonomy 173–6

Baker, A. R. H. 64, 84, 182
Balchin, W. G. V. 127, 182
Batty, M. 107, 182
bells 3, 18, 74
Bennett, R. J. 84, 107, 149, 182
Bernstein, R. J. 104, 182
Berry, B. J. L. 8, 107–8, 126, 133, 182
Betjeman, J. 74, 183
Billinge, M. 2, 81, 183
Blainey, G. 134, 183
Boddy, M. 176, 183
Bogdanor, V. 172, 183
Bordessa, R. 7, 183
Bosque Sendra, J. 166, 183
Braithwaite, R. B. 87, 183
Browett, J. 179, 183

Brunsden, D. S. 162–3, 183
Bunge, W. 7, 183
Burton, I. 129, 183
Butler, D. E. 172, 183
Butler, R. A. 74, 183
Buttimer, A. 81, 183

capitalism 25–49
Carter, H. 62, 183
Castells, M. 69, 183
chaotic conception 63, 124, 142
choice 13, 15, 16, 18, 20–1, 22
Clark, C. 123, 183
Clark, G. L. 176, 183
class 26, 28, 30, 31, 34, 63, 65, 117
Clayton, K. M. 164, 183
Coffey, W. J. 137, 183
Coleman, A. 135, 183
Collingwood, R. G. 56, 76, 87, 183
competition 10, 13, 27–8
concentration and centralization 29, 31, 32
consensus 34, 35
consumer-sovereignty 14–15, 18, 64, 147
context 21, 22, 40, 64–72
contracts 33
Cooke, R. U. 129–30, 184
cosmology 4
Cox, K. R. 132, 173, 184
Cox, N. J. 96, 106, 184
Crewe, I. 172, 184
crisis 35–6
critical theory 104–5

culture 23, 36–7, 45, 69–70, 73–4, 115
cycles 10, 32, 37

Davie, M. R. 78, 184
Davies, T. 175, 184
democracy 14–19
Detroit 12
development 6–14, 40–1, 179
doing research 62–72
Downs, A. 14, 184
Duncan, J. S. 56, 57, 174, 184
Dunleavy, P. 63, 65, 141, 184

economic base 38–43, 67, 172, 173
economism 173–6
education 5, 101–2, 119–20, 123, 144–67
Elazar, D. J. 45, 184
Eliot Hurst, M. E. 50, 53, 124, 184
Elliott, W. E. Y. 163, 184
emancipation 104–5, 115–18, 131–2, 159, 178, 179–81
empirical work 2–3, 4, 62–72, 72–82, 169–73
empiricism 3, 53–4, 57, 58, 63, 81, 83, 145, 148, 151–2, 154–5, 168, 170, 171
enterprise 8–9, 10–12, 18
environment 44, 55–6, 115
environmental determinism 19, 53–4, 125, 137, 139, 157
environmentalism 126–32
epistemology 52, 170
ethnography 60–72
Eyles, J. 3, 184

family 36–7, 42, 47
Finer, S. E. 16, 184
Fothergill, S. 7, 184
Fudge, C. 176, 183

Gale, S. 21, 184
Gamble, A. M. 16, 185

gatekeepers 164–5
gender 36
generality 91–4
Geographical Association 143, 144, 161–2, 164
geopolitics 47
Giddens, A. 53, 61, 64, 83, 84, 99, 185
Gill, D. 158, 185
Golledge, R. G. 81, 85, 185
Goodson, I. F. 123, 185
Goudie, A. S. 127, 185
Gould, P. R. 84, 122, 132, 185
Graf, W. L. 122, 185
Gregory, D. 2, 51, 57, 61, 64, 81, 84, 138, 140
Gregory, K. J. 126, 128, 185
Gregory, S. 145, 157, 185
Gregson, N. 61, 62, 185
guardian ethic 163
Guelke, L. 55, 56, 73, 75, 77, 78, 185, 186

Haggett, P. 54, 92, 126, 133, 134, ' 186
Hall, P. 76, 186
Harries, K. D. 152–3, 186
Harrison, P. 7, 186
Harrison, R. J. 4, 92, 186
Hart, J. F. 122, 139, 140, 186
Hartshorne, R. 50, 52, 186
Harvey, D. 10, 24, 26, 42, 47, 83, 85, 87, 93, 101, 117–18, 124, 135, 141, 178, 179, 181, 186
Hay, A. M. 87, 111, 186, 187
Herbert, D. T. 169, 187
hermeneutics 73
higher education 146–54
history 76–7
Huckle, J. 158, 187
human agency 39–40
humanistic geography 54–5, 92, 163–4
Husbands, C. T. 63, 65, 187

hypothesis 4, 84, 85–90, 95, 96–7,
    110, 133

idealism 55, 58, 75
ideology 11–12, 46–8, 63–4, 117,
    119, 120, 178–9
inequalities 7, 8, 13–14
innovation 10–11
Institute of British Geographers 161,
    164
instrumentalism 84
interpretation 24, 51–2, 55, 59, 67,
    68, 72, 93

Jackson, P. 69, 187
Johnston, R. J. 1, 11, 53, 54, 57, 62,
    64, 65, 69, 89, 90, 93, 98, 106,
    111, 114, 115, 120, 121, 123,
    124, 125, 126, 128, 131, 132,
    135, 140, 145, 171, 172, 187, 188
Joseph, K. 119, 158–9, 188
journals 163–5

Keat, R. 84, 99, 188
Kidron, M. 7, 188
King, L. J. 85, 188
Kirk, W. 55, 188
Knorr-Cetina, K. 147, 188
Kolakowski, L. 53, 188
Kondratieff cycle 10

Laan, A. van der 61, 188
Lacey, A. R. 53, 188
Lakatos, I. 62, 188–9
Langton, J. 139, 189
language 59, 61
law 87–8, 91, 93–4, 102–3, 105,
    109–10, 133
learned society 162–4
legitimation 34, 35, 46, 79, 101
leisure 17–19
Ley, D. 56, 57, 70, 71, 94, 174, 184,
    189
Lichtenberger, E. 166, 189

literature 74–6
locale 22–3, 64–5
locational analysis 34, 35, 46, 79, 101
Lukács, G. 132, 189

Mann, M. 45, 189
Marshall, J. U. 97, 189
Massey, D. 65, 139, 189
mechanism 24, 51, 56–7, 75, 78, 79,
    90, 93, 104, 116
Mercer, D. C. 1, 53, 115, 168, 189
methodology 4, 52, 62–72
Miller, W. L. 65, 189
mode of production 25–32
money 25–7, 32
Morrill, R. L. 92, 189

nature 131–2
neighbourhood effect 15, 42
Nystuen, J. D. 133, 189

Oakeshott, M. 50, 73, 76, 190
O'Loughlin, J. 179, 190
Olsson, G. 103, 190
ontology 52
Openshaw, S. 98, 190
opting out 12–14
O'Riordan, T. 115, 131, 164, 190
Orme, A. R. 127, 190
overproduction 29

paradigm 1, 2, 15, 62, 77
Park, R. E. 122, 190
Peet, R. 168, 190
Pepper, D. 131, 190
phenomenology 55–6, 69, 124, 135
Philbrick, A. K. 133, 190
Phillips, D. R. 169–70, 190
philosophy 4, 51, 52–8
physical geography 5, 106–7, 122,
    125, 126–32, 141, 142, 145, 150,
    157
Pickles, J. 55, 72, 77, 98, 124, 190

place 38–43, 45–8, 64–5, 67, 93, 138–40, 155
planning 107–8, 145–6
Pocock, D. C. D. 74, 190
policy analysis 108–9
political parties 14–16
Popper, K. R. 76, 97, 131
positivism 3, 4, 53–4, 57–8, 64–8, 83–5, 87, 88, 94, 95, 99, 102–3, 105–6, 109, 127, 148, 150, 154–5
power 31, 46–8
Pred, A. R. 40, 65, 190
professionalism 153, 160–1
profitability 27–8, 30, 32
pseudopositivism 107, 110

quantification 4, 84–5, 95–9, 102, 126, 133, 134, 149–50, 166
questions 170–1

Ramphal, S. S. 71, 144, 191
rationality 35–6
realism 4, 58–62, 80–1, 88, 92, 104–5, 127, 170–1, 172, 178–9
recreation 17–19
regional geography 5, 54, 137–40, 145, 155, 158
relevance 4–5, 100–2
religion 36, 60
resource analysis 128, 129
Robertson, D. 14, 191
Rose, C. 73, 191
Royal Geographical Society 123, 144, 161–2, 164
Rushton, G. 133, 191

Sack, R. D. 84, 134, 135, 191
sampling 96–7
San Francisco 69–70
Sayer, A. 63, 65, 69, 79, 84, 95, 128–9, 131, 134, 191
scale 67, 71, 158, 176
Scarbrough, E. 62, 63, 191

Schaefer, F. K. 54, 133, 191
Schattschneider, E. E. 15, 191
school education 154–9
science 5, 85–90, 90–5, 133
scientific method 4
Sennett, R. 93, 112, 113–14, 191
Sheffield 3, 17, 71, 78–9, 81, 116
Short, J. R. 67, 120, 191
Silk, J. 121, 191
Simmons, I. G. 106, 191
singularity 64, 91–4
Smith, D. M. 100, 191
Smith, N. 179, 191
South Africa 8–9, 10, 75
Souza, A. de 140, 191
sovereignty 35
spatial fetishism 134
spatial science 5, 54, 84, 92, 125, 132–7, 142, 145–6, 151–2, 172
state 32–6, 45–8, 78–9, 101, 175
stereotypes 71–2, 77, 90–1, 112–13
Stoddart, D. R. 100, 123, 192
story-telling 72–81, 87
structuralism 2–32, 56–7, 58, 75
structuration 61–2, 72
superstructure 36–7, 43–5, 60, 67, 103, 111, 172, 173
Sutherland, J. 74, 192
Swindon 3, 71
systematic geography 5, 140–1

Taylor, P. J. 67, 69, 98, 99, 118, 123, 124, 158, 170, 172, 192
technical control 102–3, 105–12
territoriality 135–6
text 73, 76, 80
textbooks 1, 3, 165–6
theory 79–80, 169–74
Thomas, K. 131, 192
Thrift, N. J. 57, 114, 192
time-geography 65–6
Trow, M. 147, 192

underconsumption 29

underdevelopment 9–10, 12, 40–1
uniqueness 64, 91–4
USA 2, 12, 15

values 19, 81
Vancouver 7
voluntarism 57, 61, 63, 64
voting 6, 14–17, 65, 77, 89, 96, 98,
    171, 172

Wallerstein, I. 121, 192
Williams, R. 73–5, 192
Williamson, J. G. 8, 192

Wilson, A. G. 107, 192
Wirth, E. 84, 192
Wise, M. J. 100, 193
Woldenberg, M. J. 126, 193
Wooldridge, S. W. 125, 193
World Three 2, 3
Worsley, P. 130, 150, 157, 193
Wrigley, N. 99, 193

Yeates, M. H. 62, 193

Zelinsky 101, 193